FIRE ENGINES
IN NORTH AMERICA

TRIAL OF ALL THE STEAM ENGINES OF CINCINNATI ON THE OCCASION OF THE OPENING OF THE OHIO AND MISSISSIPPI RAILROAD.

A demonstration of all the steam engines in Cincinnati on the occasion of the opening of the Ohio and Mississippi railroad. Prominently displayed at the front of this print is a Latta steam engine.

a division of Book Sales, Inc.
110 Enterprise Avenue
Secaucus, NJ 07094

Publishing Director: Frank Oppel
Composition: Meadowcomp Ltd.
Design: Tony Meisel, AM Publishing Services
Origination: Regent Publishing Services Ltd.
Printing: Impresora Donneco Internacional, S.A. de C.V.

Printed in Mexico

ISBN: 1-55521-674-9

CONTENTS

FIRE ENGINES
IN NORTH AMERICA

SHEILA BUFF

THE WELLFLEET PRESS

Vignettes of New York's Metropolitan Fire Department. The Fireman's Hall building at upper left was built for the volunteers in 1854. The statue on top of the building represents the legendary Harry Howard, who was chief engineer from 1857 to 1860.

INTRODUCTION

The threat of devastating fire has been with mankind for millennia—for longer than the control of fire itself—but it is only in the past few centuries that organized and effective means to combat the flames have been available. From the earliest hand pumpers of colonial times to the massive combination rigs of today, firefighters have always sought to provide the best protection possible to the people who depend upon them.

The focus of firefighting has changed greatly over the centuries. The low-volume bucket brigades and small hand pumpers of colonial times could do little to save a fully involved building. The goals of firefighters of that period were to prevent the fire from spreading to nearby structures, and to salvage whatever possible from the flames. Out of these concerns grew that unique institution of a self-reliant and civic-minded people: the volunteer fire company.

In the golden century of the volunteer fireman (roughly from 1760 to 1860), the nature of firefighting changed. Hand pumpers grew substantially larger and could pump considerably more volume; more and more volunteer companies were organized; and municipal water supplies became more sophisticated. Much emphasis continued to be placed on containing the fire and salvaging goods and property, but now the blaze could be extinguished before the structure was consumed.

Unfortunately, another change in the volunteer service in that century was away from firefighting and toward just plain fighting. In the cities, rivalries between companies and political cronyism led to brawls and corruption. Fire protection suffered, and the call for professional fire departments, removed from the influence of politics (at least theoretically) and rowdies, became very loud. At the same time, an important technological development, the steam-powered fire engine, was perfected. As Miles Greenwood, chief of the Cincinnati fire department (the first paid department in America) said, "Steamers never get drunk. They never throw brickbats. Their only drawback is that they can't vote." By 1871 even Philadelphia, the home of the first formal volunteer fire companies in America, had moved to a paid fire department.

Technological change in the second half of the 19th century meant change for the fire service as well. Steam engines were too heavy and cumbersome to pull by hand, and horses became residents in fire halls. Improvements in aerial ladders and the development of water towers went hand in hand with taller buildings. Improved signalling meant that firefighters got the alarm sooner and more accurately.

It was the horrible tragedy of the Triangle Shirt Waist fire in New York City in 1911 that led to a major change in firefighting: prevention. The Triangle fire occurred in a garment-manufacturing sweatshop on the three top floors of a ten-story building. The building was unsprinklered, with two narrow, winding staircases, and had only a single fire escape. The factory employed some 500 poorly paid workers, almost all young immigrant women. When fire broke out, the panicked workers quickly discovered that the stairwell doors were locked; the inadequate fire escape pulled away from the wall and collapsed from the weight of the fleeing women. In all, 146 young women died, many because they leaped in desperation from the windows of the burning building.

The Triangle fire led to a storm of public protest over poor safety conditions. In response, New York state enacted tough fire-prevention laws, followed soon by other states. Stringent building codes, sprinkler systems, standpipes, portable fire extinguishers, inspections, fire drills, exit signs and much more became the norm.

The modern era of firefighting began with the introduction of gasoline-powered fire engines in 1906; by the early 1920s all the horses had been retired. In the decades since then, fire trucks have been continuously improved. Significant advances in aerial ladders and the invention of the aerial platform in 1958 have given firefighters vastly increased abilities to combat the flames effectively and safely.

Firefighters in North America today—both professionals and volunteers—are the best equipped, best trained, and best motivated in the world. They are increasingly called upon to deal with complex problems such as hazardous materials spills and emergency medical situations, and they do so willingly and skillfully. Yet even as we come to depend upon firefighters for more and more, we are stinting in the money we allot for their services. We fail to install or maintain smoke detectors. We smoke carelessly. We allow hazardous building practices such as truss-roof construction. And people die because smoke detectors are missing or inoperative, because a smoldering cigarette ignites a bed, because fire spreads rapidly in an open attic area. Every year the cost in lives, injuries, and property damage from fire is enormous—and largely preventable, if only we listened to our firefighters.

Despite the budget cutbacks, despite the long hours and dangerous work, despite the indifference of a public that takes fire protection for granted, today's firefighters continue faithfully to do their job. This book is dedicated to them.

The old volunteer New York Fire Department was legendary even in its own day, and became more so as time added the patina of nostalgia. This print memorializing the old-timers appeared in the 1880s.

THE OLD NEW YORK FIRE DEPARTMENT CELEBRATION.—DRAWN BY SCHELL AND HOGAN.—[SEE PAGE 683.]

1-1

1-2

1-4

1-3

FIREFIGHTING IN COLONIAL TIMES

The earliest colonists in America faced perils and hardships unknown to them in the Old World. One constant danger, however, was familiar indeed: fire. If they were helpless against disease, attack, and the rigors of a new land, the colonists could fight fire, using familiar methods and equipment.

The first permanent colony in America was founded at Jamestown in Virginia in the spring of 1607. Led by Captain John Smith, the colonists struggled against malaria, malnutrition, illness, and Indian ambushes. A near-fatal blow was dealt by an accidental fire that devastated the colony in January of 1608. The flimsy wooden structures, including the fort and stockade, were destroyed, as were most of the colonists' possessions and stored foodstuffs. Left to struggle through the rest of the winter with inadequate clothing, food, and protection, the fledgling colony was reduced from 104 settlers to just 53 in the first year. Heroic efforts by Captain Smith and the timely arrival of a supply ship from England enabled the survivors to continue and eventually to thrive.

Colonization and devastating fire came together again in 1620, when the Pilgrims, America's most famous immigrants, founded their colony at Plymouth. Among the first structures they erected was a sort of meeting hall, which promptly caught fire and burnt to the ground. The Pilgrims spent the rest of the winter sheltering on board the *Mayflower*. By 1623 Plymouth colony was a thriving village. Predictably, that winter a chimney fire destroyed seven homes. The damage done led to the first building code regulations in America, enacted by Plymouth in 1627. Among other requirements, the regulations very sensibly forbade wooden chimneys and thatched roofs. Later, in 1638, the state of Massachusetts enacted the first no-smoking law. Curiously, this law banned smoking *outside*, in response to the numerous fires started by discarded cigar butts and pipe dottle.

Regulation alone cannot stop fire. On 14 January 1653 Boston experienced the first major fire in American history. No provisions for firefighting had ever been made in the town, and the citizens were completely unprepared. The lack of watchmen and an alarm system allowed the fire, which started along the waterfront, to gain a powerful start. The lack of fire-fighting equipment of any

1-1
Robert Morris, one of the signers of the Declaration of Independence, was a member of Philadelphia's Hibernia Fire Company.

1-2
Another signer of the Declaration of Independence was James Wilson of Philadelphia, who was an active volunteer in the Hand-in-Hand company.

1-3
The president of the Continental Congress, which adopted the Declaration of Independence in Philadelphia, was John Hancock. He is remembered today for his bold signature on the document. In his own time, however, he was probably better known as a merchant and fire warden in Boston.

1-4
The patron saint of the fire service in America is Benjamin Franklin. In 1736, Franklin and four friends formed the Union Fire Company, the first volunteer fire company in Philadelphia.

1-13
Volunteer firemen often had their home fire buckets decorated with their name and the name of their fire company. This example dates from 1776.

sort allowed the blaze to spread rapidly. Efforts to extinguish the fire were haphazard, disorganized, and ineffectual. All that could be done was to salvage anything possible and try to create firebreaks by pulling down buildings in the path of the flames. (Particularly in the case of thatched buildings, this could be accomplished with surprising ease using chains or heavy ropes attached to hooks. The hook was tossed over the peak of the roof or attached to the top of a wall; a few strong tugs were often all that were needed to yank off the roof and tumble down the walls. Even more substantial buildings could be brought down in this manner, although more work was required.)

When the conflagration eventually burnt itself out, the citizens of Boston took stock. Amazingly, only three deaths were caused by the blaze, but the damage was extensive and at least a third of the population was homeless and utterly destitute.

The town of Boston responded with vigor and adopted a comprehensive fire prevention law. The law required that "there shall be a ladder or ladders that shall reach to the ridge of the house, which every householder shall provide for his house," and that every individual householder "shall provide a pole of above 12 foot long, with a good large swob at the end of it, to reach the roof of his house to quench fire in case of such danger." More importantly, the law created the rudiments of a municipal fire-fighting effort. Six "good and long ladders" and four "good strong iron crooks, with chains and ropes fitted to them" were purchased and hung outside the meeting house, to be used in case of fire. A cistern was built, and a night watchman was appointed to patrol between ten at night and five in the morning. With the act of 1653 the Bostonians also had the distinction of enacting the first arson statute in America; the punishment was death.

As another direct result of the fire of 1653, Boston ordered a fire engine for the town—the first recorded mention of a fire engine in American history. The order was sent to Joseph Jenks, an iron worker located in Saugus, an early industrial center outside of Boston. What exactly this "engine" was is very uncertain. Since 1616 a type of fire engine using a single-cylinder, single-action piston pump mounted in a reservoir had been in use in Nuremberg, Germany (where it was invented) and elsewhere in Europe. The Boston engine might have been something along the lines of a Nuremberg engine. Alternatively, it might have been the simpler sort of syringe pump that was known to the Roman Empire. It may even have been nothing more than a big tub with handles, to be kept filled with water. There is no record that the "engine" was ever delivered, much less used, and it remains a mystery.

Even so, to Boston goes the distinction of owning the first fire engine in America. This apparatus (maker unknown) was imported from England in 1678, probably as a reaction to another serious fire in 1676. At a town meeting in January 1679 the selectmen announced, "In case of fire in the town where there is occasion to make use of the engine lately come from England, Thomas Atkins, carpenter is desired and doth engage to take care of the man-

1-13

1-5

1-6

1-8

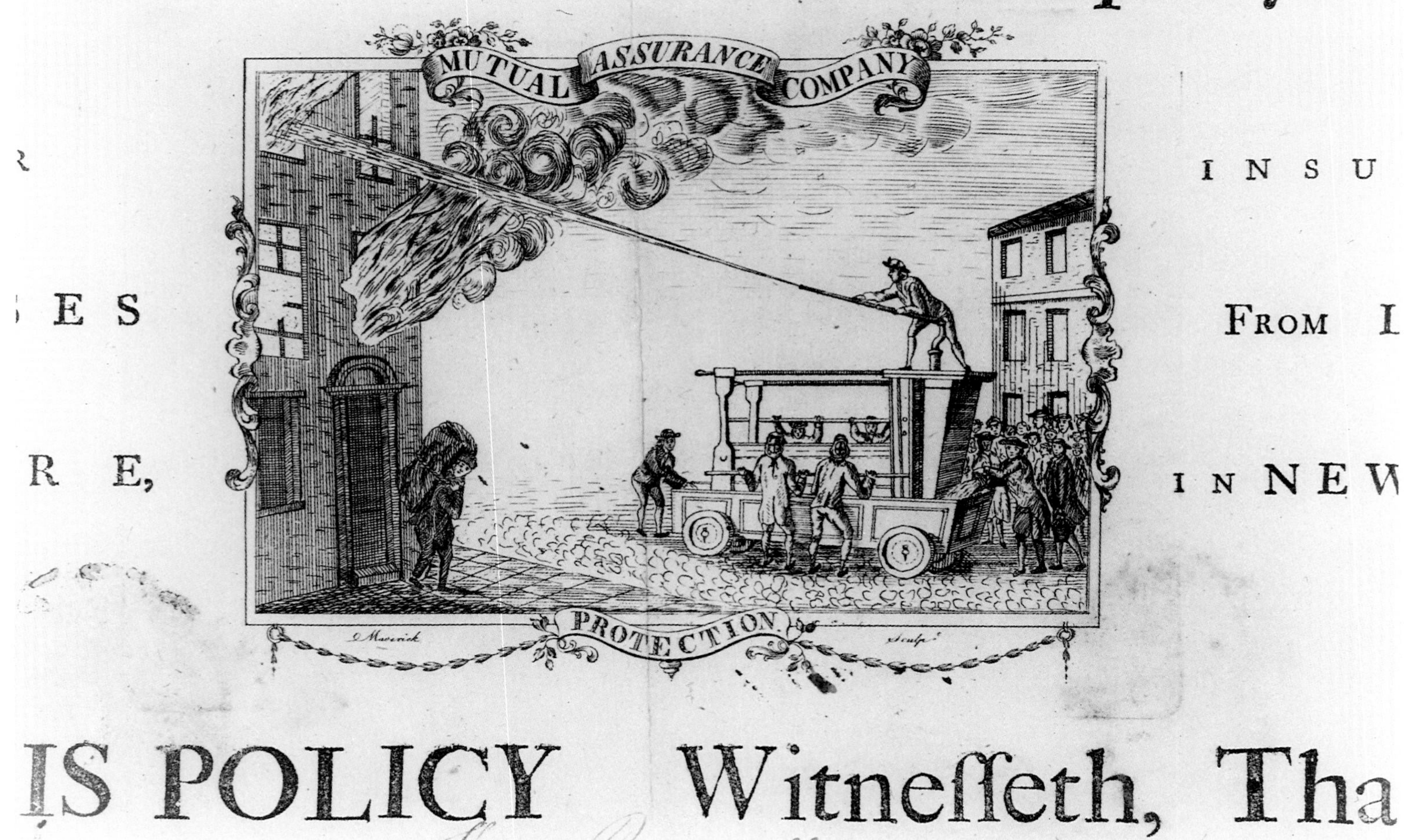

[No. 180]

Mutual Affurance Company,

R INSU

ES FROM I

R E, IN NEW

IS POLICY Witneffeth, Tha

aging of the said engine." Atkins appointed 12 assistants. Thus the purchase of the first fire engine in America led to the appointment of the first paid fire department.

The engine for which Atkins was responsible was extremely simple. A wooden box some three feet long and 18 inches wide, it rested on four wooden legs. Two men grasped the handles at the front and back and carried the engine as if it were a sedan chair. A simple single-action, single-piston pump was inside the tub, which was filled by the bucket brigade. The pump fed a thin, flexible discharge nozzle about three feet long.

Atkins stored the engine in a shed near the town jail, and it soon came to be known as "ye Engine by ye Prison." The engine seems to have had a long life. Mention of it is made in Boston town records in 1683 and again in 1702, when repairs were ordered. In 1704, the town obtained from England "one water engine suitable for the extinguishing of fire, and also brass work and other material suitable for fixing another." The idea seems to have been that by purchasing one complete engine and the hard-to-make parts of another, a second could easily be constructed in Boston. Since the rest of the engine would consist of wooden parts easily made by any skilled carpenter, this scheme made sense. It saved the town the cost of two finished engines and the considerable expense of shipping two complete engines all the way from England. The engine and parts arrived in February of 1707.

Useful as the new engines may have been for extinguishing small fires before they got out of control, Boston was nonetheless again heavily damaged by a major fire in 1711. Large parts of the town burned, destroying homes, businesses, docks, and such revered landmarks as the Old South Church. Many citizens were injured; nearly a score were dead. Out of the ashes of this conflagration rose an important development in American firefighting. Laws were enacted to divide the town into fire districts, each supervised by a fire warden. These ten men, who were to be "prudent persons of known fidelity," were equipped with badges, staffs, and the authority to organize the attack on a fire and the salvage efforts in their district.

In 1717 volunteers in Boston had organized fire societies to supplement the fire wardens and engine companies. Concerned primarily with salvage, the members of a society came to each others' aid at fires.

The primary method of fighting a fire remained the bucket brigade. Householders in Boston and other colonial cities were required to have at least one bucket for every chimney; those who practiced such hazardous occupations as baking were required to have more. Leather was the material of choice for fire buckets. In addition to being inexpensive, fire-resistant, relatively light, waterproof, and durable, leather was easily decorated. Fire buckets were often elaborately adorned with the owner's name and address, ornamented perhaps with an allegorical scene, motto, or the family crest. When the alarm of fire was sounded, the buckets were tossed out the window, to be scooped up by those already in the street running to the fire. The tossers were expected to follow on

1-5
Samuel Adams served as a fire warden in Boston in the 1760s and 1770s, and later served as a delegate to the Continental Congress.

1-6
This leather fire bucket was used by Robert Speir in his home in New York City. It dates from around 1780.

1-8
This engraving showing a fire engine in action is taken from a vignette on a policy issued by the Mutual Assurance Company of New York in the 1780s. A bucket brigade passes leather fire buckets to fill the tank, while a salvage corps member removes property from the burning building.

1-14
Allegorical symbols such as the sun or a figure of Hope often decorated leather fire buckets.

themselves, bringing along any additional buckets encountered on the way. Once at the fire scene, the bucket brigade was quickly formed into two lines. Women and children passed the empty buckets back to the water source; men passed the full buckets to the fire. All passersby were expected to join the line; those who did not might find themselves on the receiving end of a bucket of water. The two lines of men and women facing each other offered plenty of opportunity for flirtation among the younger set. To reduce this, some puritanical fire wardens ordered the lines to stand with their backs to each other. While the brigade worked to put out the flames and keep the fire from spreading, salvage teams (often members of a fire society) dragged property from the conflagration. When a home was burning, a top priority for the salvagers was the bed. In colonial times, a bed was a substantial and important piece of furniture that would often have to be dismantled to be moved. The bed key, a type of wrench used to disassemble the bed, was an essential fire-fighting tool of the time. While some worked to save the bed and whatever other furniture they could, others stuffed property into large linen salvage bags marked with their owners' name and carried them to safety.

Once the fire was extinguished, the fire buckets were dumped into a pile in a central place, where they were retrieved by their owners. This was an opportunity for enterprising boys to earn tips by returning buckets.

In 1714 the selectmen of Boston purchased a new water engine for the handsome sum of £100. Two additional engines, probably of the same make, were purchased in 1715 or so. By 1733 a total of seven fire engines protected Boston.

What sort of engines were these? The records are unfortunately silent, and none of the engines have survived. It is known, however, that letters patent for a fire engine were granted in England in 1625 to John and Roger Jones. Records in England show that a metalworker named William Burroughs improved on the Jones design and had manufactured some 60 fire engines by 1660. Several were used (with little effect) to fight the Great Fire of London in 1666. The Boston fire engines were very likely on the Jones/Burroughs model, if not actually made by Burroughs.

THE NEWSHAM ENGINE

In December of 1718 the councilmen of Philadelphia voted to purchase a fire engine from one Abraham Bickley. The cost was £50, but nothing else is known of this engine. In 1730 the council voted for a much more momentous purchase. Three fire engines were ordered from Richard Newsham, of Cloth Fair, near Smithfield in London.

Richard Newsham began life as a maker of pearl buttons. How he came to design fire engines instead is unknown, but in 1721 he had received letters patent for his new fire-engine design. He received second letters from King George II in 1725. Newsham claimed to make "the most useful and convenient

N° 13.
1803.

engines for quenching fires, which carry a constant stream with great force, and yet, at pleasure, will water gardens like small rain. . . . The largest size will go through any passage one yard wide, in complete working order, without taking off, or putting on, anything . . . . One man can quickly and easily move about the largest size in as little compass of ground as it takes up to stand in, and it is worked by hands and feet, or by hands only. Those by suction feed themselves from a canal, pond, or well, or out of their own cisterns, as opportunity offers. They are far less liable to disorder, much more durable than any extant, and play off large quantities of water, either from the engine, or a leather pipe, or pipes, of any length required (the screws all fitting each other)."

Newsham's engines were indeed the most powerful then available. They were made in six sizes. The smallest, or first size, held 30 gallons; the full amount could be discharged in one minute to a distance of 26 yards. The price without suction was £18; with suction, the price was £20. The largest, or sixth size, held 170 gallons in its cistern; the full amount was discharged in one minute to a distance to 40 yards. This top-of-the-line model cost £60 without suction and £70 with suction. (Trans-Atlantic shipment was extra.) The two smallest sizes were carried; all the larger size were on wheels. It should be noted that suction here does not necessarily mean the use of a suction hose to draft water into the tank. It could also mean a sort of trough that carried water from a pump, cistern, or other source to the tub of the engine.

The city of New York ordered two fire engines from Newsham in 1731, requesting the fourth and sixth sizes, with suction. The engines were delivered in November of 1731, and cost a total of £204 in New York currency. These new engines were put to the test almost immediately at a nasty house fire in December of that year. The engines extinguished the fire only after the house was destroyed, but it must be remembered that saving the burning building was not necessarily the first priority of a colonial fireman. Salvaging property from the building and preventing the spread of the fire to nearby structures were more important. Seen in those terms, the engines proved to be a rousing success, for they kept the fire from spreading.

1-9
One of the two Newsham fire engines imported from London to New York City in 1731, shown in a photograph taken in the late 19th century.

1-10
The oldest preserved fire engine in America is claimed to be this Newsham model, brought to Philadelphia in 1730. This photo dates to 1906.

The Newsham engines were single-acting piston pumps operated by side-stroke handles (also called brakes), by end-stroke foot treadles, or both in combination. To create a steadier flow, the pump discharged into an air chamber; a gooseneck discharge nozzle protruded from the top of the chamber. At least six men, three on a side, were needed to operate the engine brakes; two or three men were needed to operate the treadles (an experience that is oddly predictive of modern stair-climbing exercise machines). An additional bucket brigade was needed to keep the metal-lined tank or tub filled; hence, these engines came to be called hand tubs. Newsham's claim that the engine could go through a passage only a yard wide was true: the large sixth size engine was only 23 inches wide and nearly seven feet long. Cumbersome and heavy, the Newsham engines rolled on solid wooden wheels rimmed with metal, and had

1-9

1-10

1-21
The reconstructed Newsham-style fire engine at Colonial Williamsburg in operation.

to be lifted around corners. They were pulled by a square shaft with a wooden crosspiece for a handle. Despite their limitations, they were a substantial improvement over buckets. The stream could be directed (more or less) to where it was needed most, and applied steadily. Most importantly, it allowed the firefighters to do their work without having to be as close to the fire itself—although they were often still within a few yards of the flames. Although leather hose had been invented in Holland in 1672, it was primitive, leaky, and could not withstand much pressure. The Newsham engines relied solely on their gooseneck discharge pipes to direct the water stream. In practice, the foot treadles alone seem to have been rarely used for firefighting; the amount of water pressure thus generated would be far more suitable for sprinkling a garden.

FIREFIGHTING IN NEW YORK

In colonial America Boston was by far the biggest and most significant settlement, leading the way in commerce and culture. Its sheer size meant that it suffered more fires, but other towns in early America also struggled with periodic devastation by fire. For example, Charlestown, South Carolina was burned repeatedly by major fires in 1698, 1699, 1700, 1778, and 1796. Interestingly, however, the cities of New York and Philadelphia were well protected against fire from very early on.

The arrival in 1647 of irascible Peter Stuyvesant as the governor of the young colony of New Amsterdam also meant the arrival of serious fire prevention. Stuyvesant rammed a law forbidding wooden chimneys through the town council in 1648, and appointed four unpaid fire wardens. These men were to personally inspect each and every chimney in the colony; violators were fined three guilders. Fittingly, the money collected from fines was used to purchase water buckets, ladders, hooks and other paraphernalia, which were hung on buildings in strategic locations. The first fire company in New Amsterdam was formed in 1658, originally with eight young men (later expanded to 50). They patrolled the streets during the curfew hours between nine at night and five in the morning, carrying rattles that made a loud noise when twirled. Their job ostensibly was to discover fire; in fact, they acted as a sort of police patrol as well. The patrollers were not particularly popular, since they often intercepted citizens returning home late at night from places they would rather not have publicly known.

Stuyvesant took an additional step toward fire control in 1658, when 150 leather fire buckets for municipal use were purchased.

Dutch rule of New Amsterdam ended in 1664, when the English seized the colony in a surprise attack. The town was renamed New York, in honor of the Duke of York (brother of King Charles II), and life continued much as before—fires included.

Engines Nos. 1 and 2, the two Newsham fire engines purchased by New York in 1731, were a great success. Engine No. 1 was assigned to the Hudson

1-21

Engine Company, the first organized company in the city.

Recognizing the need to maintain the apparatus, the city council formally created the post of overseer of fire engines in 1733. The position carried a salary of £3 a year when first created. Later, the salary was raised to £24 per annum.

The New York General Assembly created the Volunteer Fire Department of the City of New York on 16 December 1737. This is often claimed as the first and foremost volunteer company in America. The claim for first is argued by both Boston and Philadelphia, but the claim for foremost is won without question by New York. For the next 127 years, the New York volunteers were without question the most innovative, colorful, and active in the country—a proud tradition that lives on today.

The act authorized the corporation "to appoint strong, able, discreet, honest and sober men (not exceeding more than 42 in number). . . to have the care, management, and working of said fire engines, and other tools and instruments for extinguishing of fires." The regulations and principles stated in the act later became the model for many other volunteer departments. Among other requirements, volunteers were "to take the engines and assist in [the fire's] extinguishment, and afterwards to wash the engines and preserve them in good order." Volunteers served without pay, although after a probationary period they were exempted from serving in the militia and on juries. Appointment as a volunteer was a serious commitment. Failure to appear at a fire without reasonable cause meant a fine of 12 shillings; repeated absences meant dismissal.

The city's fire engines were so useful that more were wanted. The huge expense and delay of sending to England, however, led the council to offer a fee of £50 to anyone who could make an equivalent engine in the colony. The large sum was a powerful inducement to experiment, but only one man was successful. In 1743 Thomas Lote, a cooper and boat builder, delivered New York's third fire engine. Looking much like a well pump enclosed in a box, it rolled on four crude wooden wheels. Although he followed Newsham's principles, Lote used an end-stroke design with the air chamber in the center; the discharge nozzle protruded from the top of the box enclosing the chamber. From Lote's extensive use of brass trim this engine, the first built in America, came to be known as Old Brass Backs. The proud possession of Engine Company No. 3, it was quartered on Nassau Street opposite City Hall. By the time of the Revoltionary War a number of well-known fire companies were active in New York, including the Heart to Heart, the Hand in Hand (dating at least to 1762 and probably earlier). By 1781 another well-regarded group, the Friendly Union Fire Company, was active.

FRANKLIN, PHILADELPHIA AND FIREFIGHTING

Compared to other towns in colonial America, Philadelphia was miracle of fire safety. Probably because brick construction was heavily used, the town experienced no major fires until 1730, when a fire on Fishbourne's wharf near Chestnut Street destroyed some warehouses and houses. Compared to the conflagrations that regularly ravaged Boston, this was a minor fire, but the people of Philadelphia took it very seriously. Town officials ordered 400 new fire buckets and three fire engines. Two of the engines were ordered from Richard Newsham of London; the sizes specified were the fourth and sixth. The fourth size, costing £35 without suction and £40 with, held 90 gallons in the tub, which could be discharged to a distance of 36 yards.

The order for the third engine went to Anthony Nichols, a local mechanic, in 1732. Nichols built an engine that was almost certainly very similar to the Newsham engines. Although the Philadelphians deemed Nichols's engine superior in pumping strength to the Newshams, little else is known about it. Possibly it derived its superior strength from larger size, which would have made it heavy and difficult to move through narrow streets. One of the Newsham engines from 1730 survives to this day, but the Nichols engine disappeared from history long ago. It has the distinction of being the first hand-operated fire engine built in America.

Benjamin Franklin, Philadelphia's most famous citizen, had moved to that city from Boston in 1723, when he was just 17. He was acutely aware of the dangers fire posed to any city, and constantly crusaded for better fire protection in the pages of his newspaper, the *Pennsylvania Gazette*. (It is in an article about the dangers of warming pans filled with hot coals that Franklin's famous aphorism, "An ounce of prevention is worth a pound of cure," first appeared.) Franklin took his own words to heart. On 7 December 1736, he and four friends officially formed the first volunteer fire company in Philadelphia. They called themselves the Union Fire Company, and modeled themselves along the lines of Boston's mutual-aid societies. A significant difference, however, is that the members of the Union Fire Company agreed to fight any fire, not just a fire affecting a member. The company was limited to 30 members. Among other requirements, each agreed to provide six leather fire buckets and two linen salvage bags at his own expense. Unlike the volunteer fire companies of New York, the Union Fire Company, and those that followed it, were self-governing, self-financing organizations with no relation to the municipal government.

1-11

A fire engine purchased by the Friendship Fire Company of Alexandria, Virginia, in 1755. George Washington donated another fire engine to the company in 1764.

As Franklin recounted in his *Autobiography*, the volunteer idea was so popular that other companies quickly sprang up: "The utility of this institution soon appeared, and many more desiring to be admitted than we thought convenient for one company, they were advised to form another which was accordingly done; and this went on, one new company being formed after another, till they became so numerous as to include most of the inhabitants who were men of property."

The second company to be formed was the Fellowship Fire Company, organized in 1738 by Franklin's arch-rival Andrew Bradford, the editor of the *American Weekly Mercury*. In order to be the largest fire company in the city, Bradford allowed 35 members. By 1742 there were a dozen other companies, including the elite Hand-in-Hand Company, consisting of the city's leading citizens. Other companies, such as the Hibernia, formed along occupational or ethnic lines. By 1776 there were 18 companies in the city. Franklin's Union

1-7
Times have certainly changed for firefighters. Here a reconstructed Newsham-style fire engine poses next to an American LaFrance pumper from the 1970s at Colonial Wlliamsburg.

1-7

1-12
This woodcut of "a machine for delivering persons from burning houses" appeared in the Pennsylvania Magazine *in 1775.*

1-15
An early Philadelphia-style end-stroke hand pumper made by Richard Mason in 1792. Constructed of wood and iron, this little machine stands just over four feet tall and six feet long.

1-16
Making a fire bucket the old-fashioned way, a costumed leather-worker at Colonial Williamsburg attaches the rim. The bucket will later be painted and varnished.

Fire Company remained in active service until 1820, and was finally disbanded in 1843.

Although it cannot be said with certainty that Ben Franklin invented the concept of the volunteer fire department, his enormous prestige and the obvious success of the Philadelphia fire companies certainly popularized the idea.

Firefighting owes much more to Ben Franklin. His invention of the Franklin stove improved the heating in colonial homes and also reduced the incidence of chimney fires and fires caused by flying sparks and cinders. Franklin is also responsible for another significant fire-prevention device: the lightning rod, invented in 1753.

Finally, to Franklin goes the honor of founding, in 1752, the first true fire insurance company in America, the Philadelphia Contributionship for the Insurance of Houses from Loss by Fire. In contrast to the English model, Franklin's company and those that followed did not sponsor their own fire brigades. Since the volunteer companies were pledged to put out any and all fires, not simply those affecting policy holders, the insurance companies took to making regular donations to the volunteer brigades. The funds were welcome to the fire companies, whose members had previously paid all expenses from their own pockets.

In England, insured buildings were identified by fire marks, plaques of metal carrying an emblem representing the insurance company, which were affixed to the insured buildings. This allowed the sponsored fire brigades to know immediately if the building was covered by their company. If it was, they attacked the fire and salvaged what they could; if it wasn't, they let it burn.

American fire marks also identified the insuring company, but the fire was fought no matter which insurer, if any, covered the structure. In gratitude for the volunteers' efforts, however, the insurance company concerned would often make a special bonus donation to the firemen after a building was saved. In reality, the American fire marks were little more than advertising. Today, they are desirable collectibles, both for their historic significance and their attractiveness as a form of art.

"PHILADELPHIA-STYLE" ENGINES

For decades, virtually every fire engine in America had been imported from England or from Holland, and all were built along the same side-stroke Newsham design. In Philadelphia in 1768, a volunteer fireman and mechanic named Richard Mason built a successful engine with a new, end-stroke design. Because the brakes were placed at either end of the water tank, the bucket brigade could fill it more easily. Mason offered the engines in four sizes, ranging from the smallest, needing six men to operate, to the largest, needing 14. Mason was soon receiving orders from all over the colonies, including some from as far away as the West Indies.

That Richard Mason began manufacturing fire engines in the late 1760s for well-organized and well-equipped volunteer fire companies is a reflection of an

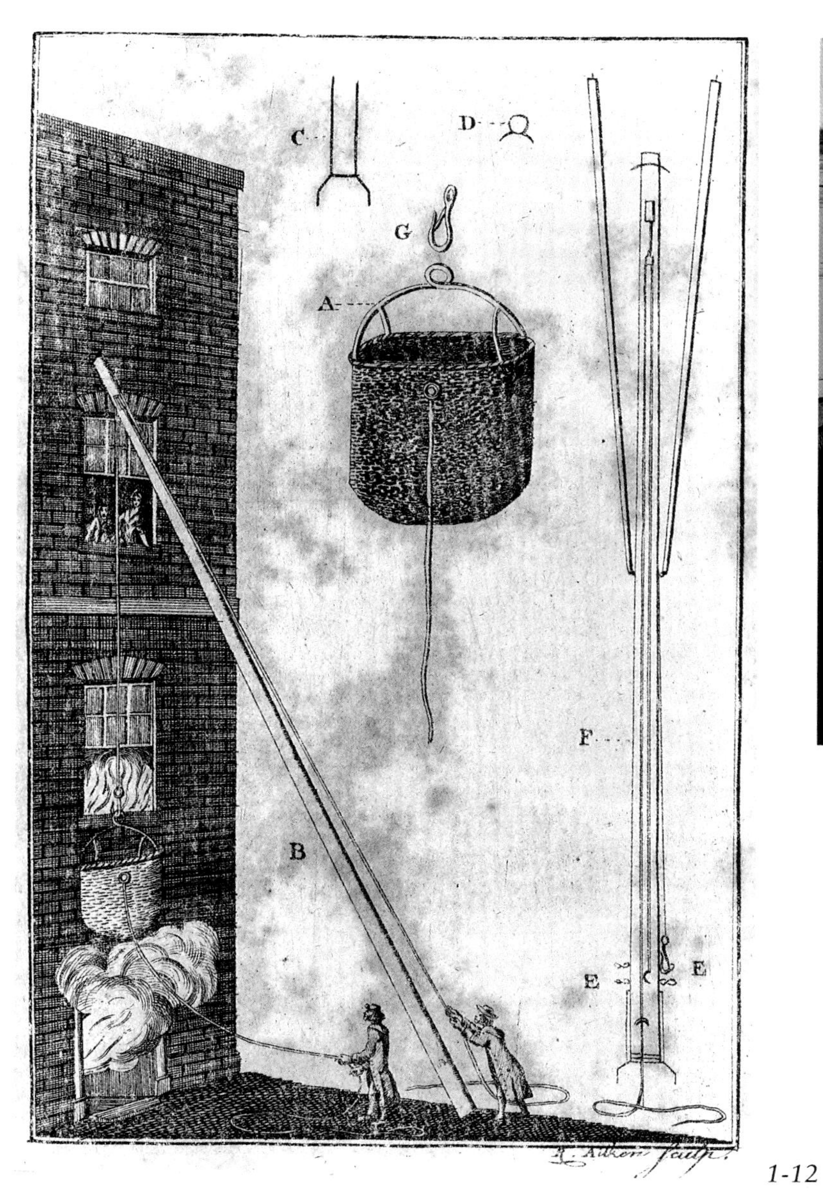

1-12

1-16

1-15

1-17

1-18

1-19

important trend in the increasingly self-reliant American colonies. By now the colonists had reached a very high level of civic, political, and economic development. No longer did they feel dependent on England; no longer would they stand to be treated as dependents. The growth and diversity of the volunteer fire companies mirrored the growth and diversity of the colonies. In every city the leading citizens were almost always also firemen. In every city there were fire companies that were organized among national, religious, and occupational groups—or combinations of all three. But no matter what their affiliation, the fire companies fought all fires, to the benefit of everyone. The voluntary organizational structure of the fire companies proved an excellent way to foster political discussion and, when the time came, action.

When the extremely unpopular Stamp Act of 1765 was enacted, for example, the stamp agent for Boston was expelled from the ranks of the Heart-in-Hand company. The Sons of Liberty, a clandestine organization formed to protest the Stamp Act (by the Boston Tea Party, among other actions), drew many members from among the volunteer fire companies.

THE VOLUNTEERS AND THE REVOLUTION

By 1750 tensions between the citizens of Boston and the British troops quartered there were high. On 5 March, the fire bell in the town hall was rung, although no fire was evident. Civilians and soldiers formed a crowd in front of the building. Taunts were exchanged, and a captain led in a troop of eight soldiers to restore order. The soldiers fired; 11 people were wounded, five fatally. The Boston Massacre claimed the lives of three firemen. As the news spread it would have particular poignancy among the volunteer firefighters, and their influence would add to the colonists' increasing resentment of the British. When the Declaration of Independence was written, numerous volunteer firemen were among the signatories. John Hancock was a fire warden in Boston, and had presented his city with its tenth fire engine in 1772. Both Samuel Adams and Paul Revere also served as fire wardens in Boston. Jāmes Wilson of Philadelphia was an active member of the Hand-in-Hand company, as were Dr. Benjamin Rush and Francis Hopkinson; Robert Morris, the financial wizard who funded the war, was a member of Philadelphia's Hibernia Fire Company, along with John Nixon and George Clymer. There is no direct evidence that George Washington was a volunteer fireman. Young men of his background routinely were volunteers, however, and tradition says that Washington served with the volunteers of Alexandria in Virginia. He was an honorary member of the Friendship Fire Company of that city, formed in the 1770s, and donated an engine to the company. Finally, the numerous contributions of Benjamin Franklin to firefighting and fire prevention need no further mention.

When the fighting began, many firemen became militiamen, even though they were exempt from military service. Their patriotism caused a dilemma in

1-17
A close-up view of the pumping mechanism of a reconstructed Newsham-style engine built at Colonial Williamsburg in the early 1980s.

1-18
The chain mechanism and pump cylinders are clearly visible in this close-up look at a reconstructed Newsham engine.

1-19
Water supply was always a problem for colonial firefighters. Here a Newsham-style engine at Colonial Williamsburg is filled using a barrel and trough--an early version of "suction." Note the leather fire buckets to the left of the engine.

1-20
Hand-sewn leather hose tended to burst under pressure. Note the overlapping seams on the hose at front and the metal attachments to the other hoses. The hoses here are being made by a craftsman at Colonial Williamsburg.

many places, since the threat of fire was all the greater during war. The burning of New York in September of 1776 illustrates the situation. The city was of critical strategic importance to the American army under George Washington. The British managed to occupy the city, however, and Washington—along with most of New York's firemen—was forced to retreat. On 20 September, several fires broke out in the city. Their origin is controversial to this day. Some feel the fires were almost certainly the work of arsonists supporting the revolutionary cause; others claim that some British sailors on shore leave started the first fire accidentally. The few trained firemen left in the city couldn't fight the fires effectively, particularly when they discovered, depending on which story is accepted, that their engines had been disabled, their hoses cut and their fire buckets holed, or that their apparatus was in poor working order due to neglect. The fire bells were gone, taken by the retreating rebels to be melted for bullets. No matter how it started, the conflagration quickly grew out of control and destroyed some 500 buildings, or approximately a quarter to a third of the city. British soldiers, believing that the rebels had deliberately fired the city, summarily arrested hundreds of citizens and hanged several on the spot. Sailors called in to help fight the fire looted homes and shops instead.

To this ravaged city the British sent thousands of troops; thousands of loyalist refugees joined them. The city quickly became unpleasantly overcrowded, a situation made worse by another serious (and definitely accidental) fire in 1778.

Many volunteer firemen served with distinction in the Revolution—and most were killed. Volunteers from New York City are known to have been at the Battle of Long Island, Valley Forge, Princeton, Trenton, and elsewhere, as well as at the final victory at Yorktown.

When victory came in 1783, the militiamen became firemen once again.

VOLUNTEERS AND THEIR MASHEENS

With the Revolutionary War behind them and a nation to build ahead of them, the citizens of the new United States of America rapidly expanded the bounds of settlement. As towns and cities grew, the demand for fire protection grew as well. New volunteer companies were formed, and the manufacturing of fire engines became a worthwhile business. Yankee ingenuity, already legendary, blossomed.

Richard Mason had been building fire engines in Philadelphia since 1768. He was joined in the business by his son Philip, who was active between 1795 and 1805. All told, the Masons built some 120 engines before they ceased operations for unknown reasons in 1806.

In competition with the Masons was another builder, Pat Lyon. A blacksmith born in England in 1769, Lyon came to Philadelphia in 1793 and had built his first fire engine by 1794. In 1799 he had refined his design and produced an immensely successful double-piston, double-deck, end-stroke hand engine. Not a modest man, Lyon supposedly offered an eternal guarantee on his engines, saying that if anyone ever made an engine better than his, he would rise from his grave and make one even better. Firemen throughout the country were convinced of the merits of a Lyon engine. Between 1794 and 1826 he sold some 150 pieces of fire apparatus. In 1802, the Good Will Fire Company of Philadelphia, established that year, purchased a Lyon fire engine and hitched a horse to it to pull it through the muddy streets. This was the first horse-drawn apparatus in the country. Lyon was involved in two other related firsts. In 1803 he built the nation's first hose wagon for the newly formed Philadelphia Hose Company No. 1. These enthusiastic volunteers topped their rig with a spring-operated clapper that rang a bell as the wheels turned—the first fire apparatus bell in America.

Lyon's business was not confined to fire engines. Pumps of various sorts were needed for other uses, such as removing water from the bilges of ships and watering gardens. Lyon supplied them, along with printing presses, fireproof iron chests, and replacement parts for other fire engines. He was apparently a hard-headed businessman, for one of his newspaper advertisements

2-22
This elaborately decorated hose wagon was once the property of Marion Hose Company No. 2. It is now lovingly preserved in the Fire Museum of Maryland.

2-23
The Carlstadt, New Jersey, Fire Department owns this handsome 1865 James Smith side-stroke pumper.

states, "Orders from any Part of the Continent executed with Punctuality, for *ready Money* only."

Lyon's basic design was soon copied by other builders, and the rigs came to be called "Philadelphia-style" engines. Foreign engines continued to be imported, but a number of apparatus makers had become well established by the first decades of the 19th century. In Philadelphia, Lyon faced competition from James Sellers, who devised a cumbersome and inefficient engine called a hydraulion. This machine combined an end-stroke pump with a hose reel—a surprisingly modern concept that was nonetheless ineffective. Sellers managed to sell seven of these machines to the Philadelphia fire brigade, but they never became popular.

More successful were the engines built by Jacob Perkins, although very little is known of them. Perkins himself was born in Newburyport, Massachusetts, in 1766. By 1801 he had patented a new type of compact and efficient pump. It was not until 1812, however, that Perkins managed to sell a fire engine using his pump to the town of Newburyport. By 1817, Perkins had moved to Philadelphia and gone into partnership with Thomas P. Jones. The firm of Perkins & Jones Patent Fire Engine & Hose Pump Makers produced several sizes and types of fire engines, as well as "domestic engines" for watering gardens, washing windows, and the like. The Perkins & Jones village engine was designed to operate with eight men and raise enough pressure to play a stream over a four-story house. It cost $250. The largest Perkins & Jones engine discharged 1 1/2 tons of water per minute; the maximum stream height was 180 feet through a nozzle of 1 1/8 inches. The cost was $1,000; the copper suction pipe was an extra $50. By 1819 Perkins was no longer in partnership with Jones, but the business was apparently successful. Perkins claims to have sold more than 200 engines in two years. In 1819 Perkins moved to England, where he was honored for his work in fire control but found no buyers for his engines. The business in Philadelphia was carried on by his son-in-law, and vanishes from the record in the 1820s.

In New York City, Thomas Lote had built the first American-made fire engine in 1743. Several crude, heavy hand engines were built by one Jacob Broome in the 1780s, probably along the typical Newsham lines. The wooden tank was about eight feet long, three feet wide and two-and-a-half feet deep; it held about 180 gallons of water. The condensing chamber rose from the middle of the tank. A handle was at either end, each worked by four men. By the time these machines began to wear out several years later, Broome was apparently no longer in the fire business. Very similar replacements were purchased from Abel Hardenbrooke. These engines had rear wheels that were larger than the front wheels and used side-stroke operation. The design used came to be known as the "New York style."

The late 1700s and early 1800s were a time of interesting experiments in hand engines. Rotary-type fire engines were developed. The "coffee mill" type of rotary engine was operated by turning side-mounted cranks. The "cider

MARION HOSE
2
CARLSTADT FIRE DEPARTMENT
AND ASSISTING
1881
EAST RUTHERFORD · RUTHERFORD · WOOD-RIDGE
1920
JAM ENVELOPE & PAPER
Discount Outlet

2-22

2-23

2-1

2-4

2-2

2-3

2-5

2-6

mill" type was operated by a sort of capstan or windlass; the firemen pushed on poles as they walked in a circle. Although a crank is theoretically a very efficient way of transmitting energy, rotary fire engines never really caught on. In part, they simply couldn't generate enough pumping power. The cider mill type sounds positively dangerous; volunteers could easily slip and fall as they pushed the capstan bars.

One particular rotary engine did enjoy a great deal of success and fame. In 1800 a group of New York City volunteers purchased a large coffee-mill rotary engine and mounted it on a scow docked on the East River. Known as the Floating Engine, this was America's first fireboat. The 12 volunteers would row it to fires. This primitive fireboat remained in operation for nearly a quarter of a century. When it was finally superseded by better designs, the rotary engine was rescued for another use by the Supply Engine Company. It was permanently mounted in a wooden shed over the deepest well in the city, from whence it pumped water to pumpers fighting fires in the area. This company enjoyed the unusual distinction of never leaving its engine house.

GOOSENECKS AND CRANENECKS

The early Newsham hand engines had some serious design flaws. Among other things, the entire engine had to be turned to aim the discharge nozzle. This was a particular problem in narrow alleys and other places where turning the machine to aim the flow would mean that the brakes couldn't be operated effectively. A solution was the "gooseneck" engine—a curved discharge pipe leading from the top of the air dome and out the top of the enclosing condenser case. At the end of the pipe was a swivel connection for a long nozzle; the nozzle could thus easily be aimed at the fire. In addition, the firemen could stay that much further away from the flames and any falling debris. Although no one knows for sure exactly who originated the gooseneck engine, James Smith of New York City is the one who made it practical. Starting in 1810 and ending in 1864, Smith's firm produced some 500 engines, many of which have been lovingly preserved to this day. In the 1820s Smith faced some competition from another New York builder named Harry Ludlum. Ludlum stayed in business until about 1860, but built fewer than 100 engines.

More serious competition came from Boston. In the 1790s a coppersmith named William C. Hunneman (a former apprentice to Paul Revere) started to make fire engines that quickly became famous for their light weight and easy maneuverability. An interesting feature of these engines is that the brakes were mounted as if for side-stroke operation; in actuality, they swiveled around and the engine was operated as an end-stroke. The Hunneman firm quickly outgrew its space in Boston and moved to nearby Roxbury. William brought his sons Samuel and William Jr. into the business, which carried on for more than 90 years. In that time, the firm made slightly over 700 hand fire engines, the most of any American manufacturer. In addition, Hunneman made hose reels,

2-1
The first fire engine ever used in Brooklyn, New York, was this one, built by Jacob Broome in 1785. This lithograph of it dates to 1863. Note the liquor cask by the wheel; firemen of the time often poured liquor into their boots to keep their feet from freezing in the winter.

2-2
A hand-drawn hose carriage used by the Nereid Engine Company No. 1 of Wakefield, New York.

2-3
Firefighting tends to run in families. This hand-drawn hose carriage was used by the Exempt Firemen's Sons Company of Morrisania (now the Bronx), New York. Firemen in volunteer days were exempt from militia and jury duty.

2-4
Laurel Hose Reel No. 1, of the York, Pennsylvania Fire Department.

2-5
The hand-drawn fire engine of the Lady Washington Engine Company No. 1, of Morrisania, New York. The stuffed rooster mounted on top was a company emblem.

2-6
The Nereid Engine Company of Wakefield, New York, used this crane-necked side-stroke pumper. Note the unusually long strainer on the suction hose.

2-24
The setting and the uniforms may be different, but the idea is the same: grab the rope and pull. Firefighters from Carlstadt, New Jersey display their 1865 James Smith hand engine at a parade in 1990 celebrating the 125th anniversary of the Fire Department of New York.

2-25
The drag rope used to pull hand-drawn apparatus was often stowed on a reel between the front wheels, as on this 1865 James Smith hand engine.

ladder trucks, and other apparatus. Moving with the times, the firm later made steam engines.

Another Boston builder, Ephraim Thayer, entered the market at about the same time as Hunneman. Working with his son Stephen, Thayer began building engines in the late 1790s. His design was quite similar to Hunneman's, but his firm was not as long-lived. Stephen Thayer produced engines in the 1830s and 1840s, but by 1860 the firm was defunct. Both Hunneman and Thayer were known for building extremely powerful engines.

THE INVENTION OF HOSE

Hoses of various sorts are so critical to firefighting that it is hard to imagine a time when they did not exist. However, the first working fire hose was invented in Holland in 1672 by Jan van der Heyde. Made of leather or heavy sailcloth, these early hoses had a single sewn seam. They were made in 50-foot lengths, with brass screw threads at both ends to connect them. Conceptually sound, sewn hoses were impractical in operation, because the seam tended to leak or even burst under pressure; not much water actually got to the pump. Although hose continued to be used in firefighting, throughout the 1700s and into the early 1800s, brass or copper/brass nozzles attached to discharge pipe were used to get water onto the fire. The nozzles were often six or seven feet long; the opening was small, generally with a diameter of one to one-and-a-half inches.

The first hose company in America was formed in Philadelphia in 1803. Founded by Reuben Haines and 20 young Quaker men, Philadelphia Hose Company No. 1 also commissioned the first American hose wagon, built for them by Pat Lyon. The company saw its first official action at a stable fire in March of 1804.

In 1808 a crucial innovation in hose (in those days also spelled hoase, hause, and hoose) design was made. Instead of thread, which gave way under pressure and also rotted, copper rivets were used. The first manufacturers of riveted hose were James Sellers and A.L. Pennock, in partnership in Philadelphia. (In an early example of vertical integration, the firm of Sellers & Pennock later began making fire engines.)

Leather fire hose was made from heavy cowhide (or buffalo hide in the West). A 50-foot length could weigh over 80 pounds when dry. The leather needed to be kept greased to prevent cracking. The recommended method was to melt beef tallow with neat's-foot oil. The resulting odoriferous substance, known as slush, was rubbed into the hose, presumably by the company member with the least seniority. The aroma of well-slushed hose in the engine house on a hot day must have been memorable.

The invention of practical hose changed both fire engines and firefighting techniques. The engines began to carry a reel of hose mounted above the tub. Practical suction hose was developed, allowing the firemen to draft water from

APEX TECHNICAL SCHOOL
APEX TECHNICAL SCHOOL
ONE WAY
ONE WAY
MACK

2-24

2-25

a pond, cistern, river, or the like, and bucket brigades gradually became a quaint memory in cities with reliable water supplies. No more buckets meant that fewer men were needed to battle the blaze; the confusion at the fire scene was sharply reduced. The firefighters could also get closer to the flames and even enter buildings without endangering their engines.

The invention of reliable suction hose led to another innovation. To get water on a fire distant from a water source, a number of engines would form in line. The one near the water source would pump from it, using suction hose, and fill the tank of the next engine in line. The water was passed down the line in this manner until it reached the fire. Pumping in line became a source of pride and competition to the volunteer companies. If one engine pumped its water into the next faster than the second engine could pump it out, the second tank would overflow. This humiliating event was known as "being washed," and was to be avoided at all costs. Despite the occasional brawl, pumping in line was generally efficient. Instances of 20 or 30 engines in line, stretching for more than a mile, were recorded. Suction hose quickly became standard. All the engines in New York City had been adapted to suction by 1819. The first rig specifically manufactured with suction gear was made by Sellers & Pennock in 1822.

2-7

2-7
An end-stroke pumper built by John Rogers of Baltimore for the Columbia Fire Engine Company of Washington, D.C., in a photo dating from 1859. The banner hanging on the pumper would suggest that it was already something of a parade piece.

2-8

2-8
In this rare action photo, two hose teams race each other in Deadwood, Dakota Territory, to celebrate the Fourth of July in 1888. As this picture clearly shows, buildings in frontier towns tended to be flimsy wooden structures that burned easily and often.

2-9

2-9
The first fire engine of Hempstead, New York, dates to 1832. Before being moved to Hempstead, this machine served as Engine No. 2 in the Brooklyn volunteer fire department.

The fire engine itself could carry only a limited amount of hose. Hose wagons and carts were developed to haul additional lengths. In 1819 David Hubbs, of New York's Eagle Company No. 13, invented a lightweight, fast, two-wheeled cart for carrying hose. Basically a hose reel mounted on an axle between tall, thin wheels, the cart and those after it became known as "Hubbs' Babies." The cart could be attached by hooks to the back of the engine, or it could be pulled separately by two or more firemen. Carts of this sort were also known as tenders and leader carts, but were popularly called jumpers; lightweight four-wheeled hose carts were known as spiders, and the heavier four-wheelers were called crabs. A typical jumper carried some 600 feet of hose; a spider could carry 1,000. Hose carts reached a fairly high level of mechanical sophistication. The four-wheeled hose cart of Columbian Hose Co. No. 9 (Silver Nine) of New York, for example, carried 1,000 feet of hose and weighed in at some 1,200 pounds. A winch was used to turn a series of gears in the reel drum to roll in the hose.

Hubbs' innovation led to the formation of numerous hose companies. It could also be argued that the prestige of being a nozzleman was greater than being an anonymous pumper. For whatever reason, in 1823, 19 hose companies were reported in Philadelphia alone. Through the volunteer period in New York City (ending in 1865), there were 63 hose companies, compared to 57 engine companies and 18 hook and ladder companies.

INNOVATION AND PROGRESS

While new ideas and equipment were adopted by engine and hose companies, the less popular hook and ladder companies were hampered by insurmountable problems of weight and design. The wooden ladders had to be kept short so that the ladder truck could go around the corners of narrow, crowded streets. The firemen came up with a partial solution when they attached the rear wheels to the axle using a "fifth wheel" or circle iron; a projecting strap of iron, or tiller, allowed the rear wheels to be steered around curves and corners. As ladders and wagons grew longer, a steering wheel and seat for a tillerman was added above the wheels. For better control while in motion, longer ladder trucks often had handles along the sides. While some volunteers pulled the truck from the front, others grasped the handles and ran alongside. Wooden ladders were heavy, clumsy, and combustible. No ladder could be more than about 75 feet long and still be raised by hand, yet buildings were growing taller. A better solution—aerial ladders—would not be found until the 1870s.

In the meantime, progress continued on hand pumpers. John Agnew of Philadelphia began building fire engines in 1823. His engines were famous for being extremely powerful. Agnew engines were big, especially in comparison to the older pumpers. A second-class Agnew used 8 1/2-inch cylinders and had a 9-inch stroke. All told, Agnew built about 150 engines. In 1841 Agnew began producing a new model. Built in the Philadelphia style, these engines were double-decked end-strokers, with two sets of pump brakes. Two sets of firefighters operated each handle, one standing on the ground and the other on narrow strips of decking above the water box.

Another famed manufacturer of hand engines was the Button Fire Engine Company. Based in Waterford, New York, this firm was established in 1834 by Lysander Button and his son Theodore. The workshop was conveniently located on the Erie Canal, handy for testing the pumps and shipping the finished product. By the time the firm became part of the American Fire Engine Company in 1891, some 500 Button hand engines in seven different sizes had been built. The largest needed 60 men to operate. In addition to hand engines, the firm also made two- and four-wheeled hose carts. In 1838 Button began making a type of side-stroke fire engine with a flat deck and larger water box. Because it had a fanciful resemblance to an upright piano, the design came to be known as the piano or piano box style. Lysander Button also perfected the permanently attached suction hose in 1848. The hose was attached to the inlet opening at the rear of the engine and was curved up over the top when not in use. The curved portion of the suction hose was often inserted into a large brass tube held by brackets to the top of the engine. The style was immediately dubbed squirrel-tail by the firefighters. Because the suction was preconnected, squirrel tails saved precious moments when the engine arrived at the fire. Button's squirrel-tail engines were light and handled better than the larger double-deckers.

2-26
An up-close look at some details of the Carlstadt, New Jersey department's well-preserved 1865 hand pumper.

2-27
The Diligent Fire Company of Philadelphia ran this elegant end-stroke pumper, built by Pat Lyons in 1820. This illustration dates from 1832.

2-26

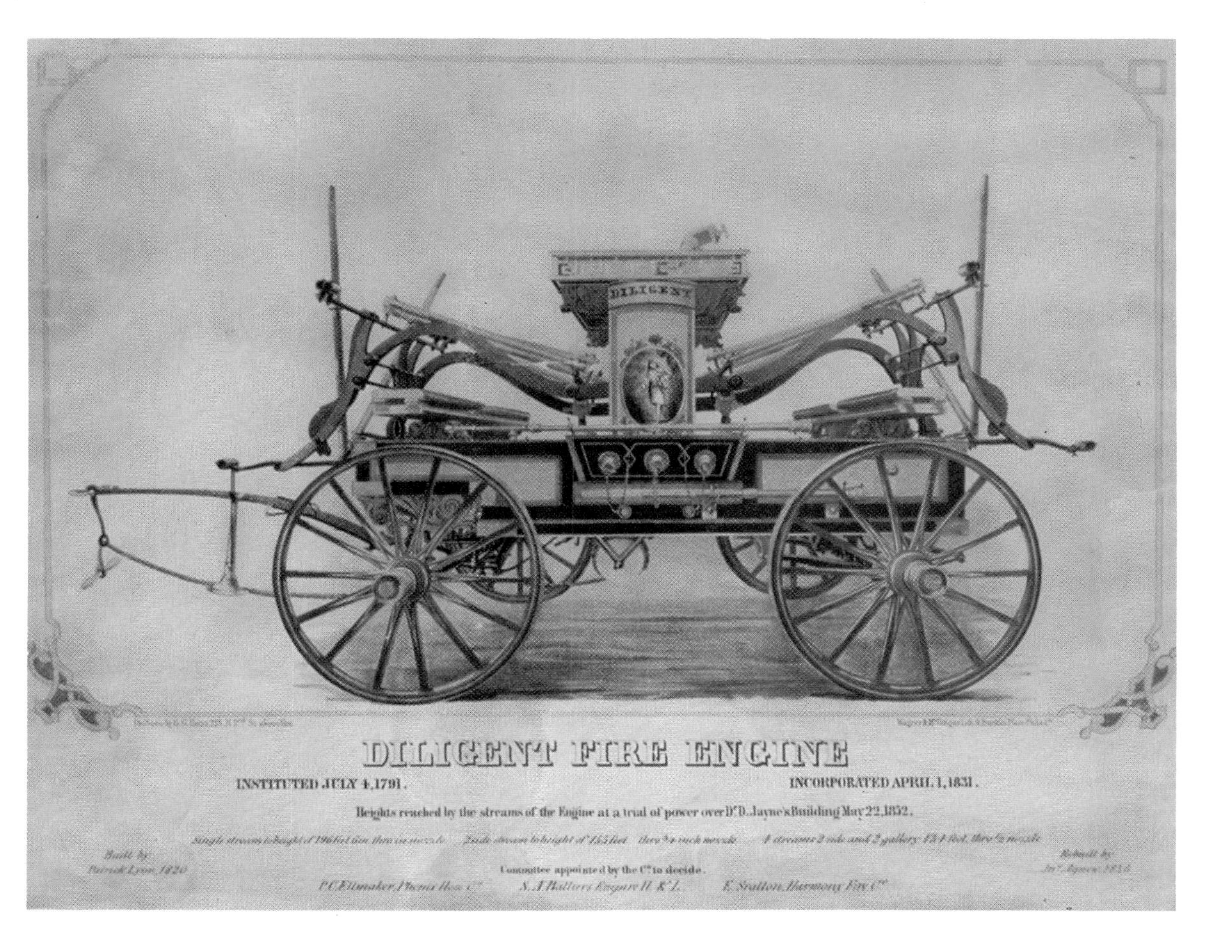
DILIGENT
DILIGENT FIRE ENGINE
INSTITUTED JULY 4, 1791.
INCORPORATED APRIL 1, 1831.
Heights reached by the streams of the Engine at a trial of power over Dr. D. Jayne's Building May 22, 1852.
Built by
Patrick Lyon, 1820
Committee appointed by the Co. to decide.
E. Stratton, Harmony Fire Co.

2-27

2-10

2-11

2-12

2-13

2-14

2-15

Button was responsible for several other important developments. The firm was the first to offer a pump with valves set at an angle of about 40 degrees. The water in these pumps traveled on one level between the inlet and the outlet, improving efficiency. To Button also goes the credit for inventing adjustable leverage in 1842. A slotted crossbar allowed the leverage on the pump handles to be changed without changing the stroke. By shortening or lengthening the leverage, the amount of water pumped out was decreased or increased. The men working the brakes, however, could maintain a steady rhythm to their accustomed cadences.

The invention of the crane-neck has been variously credited to Hunneman, Thayer, and Button in the late 1840s. By curving the main shaft of the fire engine upward and then down again (like the neck of a crane), the front wheels could be made to turn under the engine. This important improvement led to much better maneuverability, particularly when taking the engine around tight corners. In contrast, machines without the crane-neck would be forced to back at least once to get around a narrow turn.

Double-decker engines were clumsy but powerful, and power was what many fire companies wanted. In the 1840s, the aptly named Henry Waterman of Hudson, New York was commissioned by Empire Engine Co. No. 42 to build the most powerful engine he could. Waterman came through with a hand engine so powerful that it could beat the stream height and distance of an early steam engine. Waterman used a side-stroke, double-deck design with 9 1/2-inch cylinders and a 9-inch stroke. The double brakes folded up above the water box, giving the machine the appearance of a hay rick. The Empire boys at first nicknamed their machine the Haywagon. When they realized how heavy and hard to pump it was, they gave it a second nickname: Mankiller. So huge and clumsy was this engine that it was placed in reserve for use only at the worst fires. The haywagon style, however, was practical, and was adopted by other manufacturers.

Another interesting experiment of the late 1840s was conducted by Dudley L. Farnham and Franklin Ransom of New York City. They invented the rowing engine. Instead of pump handles placed at the sides or ends, the Farnham and Ransom engine had an apparatus resembling a rowing machine arranged along the top of the engine. Depending on the size of the engine, six to ten firemen were seated in pairs; they "rowed" on the "oars" (really cross-brakes attached to the extended arms of a single-acting piston) to pump the engine. The design found no favor in the inventors' home town, but some were sold to Cincinnati and some other municipalities in the Midwest.

A number of other apparatus makers set up shop in the late 1830s and early 1840s. In New York City were Pine & Hartshorn, John H. Sickels, W.H. Torboss, and A. Vanness. John Rodgers of Baltimore and Jeffers & Company of Pawtucket, Rhode Island, made New York style engines. In New York State, Seneca Falls, on the Erie Canal near Elmira in the Finger Lakes region, became a center of fire apparatus manufacturing. Because the town was already estab-

2-10
The Hornett, an end-stroke fire engine of the Goodwill Company of York, Pennsylvania, dates to 1839.

2-11
Firemen in Boston clean their machine in a plate dating to 1851. Note the two-wheeled hose reel, or jumper, at the left. The reel would be attached to the rear of the fire engine, where it would bounce or "jump" as the assembly was pulled through the cobblestoned streets.

2-12
Fire Engine No. 38 of New York City, in a wood engraving from 1854. Compare the large size of this hand pumper to those from the colonial period.

2-13
This interesting advertising poster is for Remington fire engines. Operating along the coffee-mill principle, this pump is powered by horses, presumably the same ones that drew it to the fire. The idea never caught on.

2-14
The Protection Engine Company No. 5 of Melrose, New York (now part of the Bronx) used this fire engine. It was built by James Smith in 1857 and remained in use until 1874.

2-15
A famed series of color lithographs issued by Currier & Ives in the 1850s depicted scenes from the life of a fireman. This litho is titled "The Night Alarm: Start Her Lively, Boys." It shows the Excelsior Engine Company No. 2, based on Henry Street in New York City. The man with the lantern is thought to be Nathaniel Currier, himself a volunteer fireman.

2-28
In this illustration from the 1830s, the Hibernia Engine Company No. 1 of Philadelphia is depicted with its elaborate Pat Lyons hand engine.

lished as a center for pump manufacturing, the step to fire engines was an easy one. The firm of Paine & Caldwell moved there in 1839. Silsby Manufacturing was founded there in 1845. Rumsey Manufacturing, founded in 1840, moved to Seneca Falls in 1865. Silsby would become part of the Fire Extinguisher Manufacturing Company in 1866; this firm in turn became part of the American Fire Engine Company of Elmira in 1891. Rumsey became part of the American Fire Engine Company's successor firm, International Fire Engine Company, in 1900. Thus both Silsby and Rumsey are part of the ancestry of the modern-day firm American LaFrance.

The Ahrens Manufacturing Company of Cincinnati was founded in 1851 to make hand pumpers; just a year later, the first successful steam-powered fire engine was demonstrated in Cincinnati by Moses Latta. The firm adapted quickly to the changing technology and went on to be a leading apparatus maker for a century longer.

The technology of the hand engine may have reached its highest level in the "Shanghai" engines built by James Smith in the late 1850s. This first-class, double-decker, end-stroke engine had two sets of brakes that operated alternately. Other double-deck designs were operated by the men on the upper deck on one side pushing up while their counterparts on the ground pushed down. In alternate action, the work was coordinated by levels, not sides. The men on the ground on either side worked together, as did the men on the deck.

In the heyday of the hand-operated fire engine there were no formal standards. Gradually, however, an approximate set of standard sizes and capacities evolved:

- First size or No. 1 engines had cylinders 9 to ten inches in diameter. The discharge rate was roughly 2 1/2 gallons per stroke. At least 12 men were needed to operate the engine, with room for up to 60 at a time.
- Second size or No. 2 engines had cylinders 8 1/2 to 9 1/2 inches in diameter. The discharge rate was a bit more than 2 gallons per stroke. At least eight men were needed to operate the engine, with room for up to 52 at a time.
- Third size or No. 3 engines had cylinders 7 to 9 inches in diameter. The discharge rate was just under 2 gallons per stroke. At least eight men were needed to operate the engine, with room for up to 36 at a time.
- Fourth size or No. 5 engines had cylinders ranging anywhere from 4 1/2 to 9 inches in diameter. The discharge rate was a gallon per stroke. At least six men were needed to operate the engine, with room for up to 18 at a time.
- Fifth size or No. 5 engines had cylinders of 4 to 7 1/4 inches in diameter. The discharge rate was bit more than half a gallon per stroke. At least four men were needed to operate the engine, with room for up to 16 at a time.

The stroke of the piston rods varied from manufacturer to manufacturer, ranging from eight to 18 inches. The bigger the engine the longer the pump handles; these could be from 16 to 25 feet long.

Operating a hand engine was an arduous task. After hauling the engine (the Philadelphia-style double-deck machine of New York's Americus Engine Co.

2-28

No. 6, known as Big Six, weighed over two tons) to the fire at a dead run, often for half a mile or more, the firemen would pump the brakes at a normal rate of 60 strokes a minute (a stroke was one complete up and down motion of the handle). When more water was needed, the pace was speeded up, and could reach 120 strokes a minute. At least one case of 170 strokes in a minute has been reported. At the usual 60 strokes a minute an individual fireman could pump for about ten minutes before needing to be relieved. At faster paces, the firemen tired much more quickly, being able to pump for only two or three minutes at a time. At the highest pumping rates, the hand engines could produce impressively powerful streams, as shown by records set at modern musters using preserved equipment. An 1852 second-size engine built by E. Lesley set a record of 262 feet 5 3/4 inches at a muster in 1975. The record for first-size engines is claimed by several veterans: an 1872 Button & Son at 285 1/2 feet in 1903; an 1860 Cowing & Co. at 281 feet 7 inches in 1860; an 1865 William Jeffers at 281 feet, 7 1/2 inches in 1965; and an 1895 Gleason & Bailey at 279 feet 1/2 inch in 1975.

The value of drill showed when hauling the engine and pumping it. For the most efficient handling, the men pulling the engine needed to run in step. Anyone who fell while hauling would be trampled and possibly run over by the engine, with sometimes fatal consequences. In addition, pedestrians, vehicles, and stray animals like pigs and cows were constant obstructions. The fire company had to practice sudden swerves and stops to avoid injury to themselves and their rig. Because the roads of the period were often clogged with traffic and thick with mud and ruts, the firemen would sometimes "jump" the engine onto the sidewalk, to the consternation of strollers and shopkeepers. Serious accidents could result, and the practice was frowned upon by the authorities. The firemen were capable of amazing feats. In one contest, a 16-man hose company pulled its fully equipped carriage 500 yards in just one minute, 23 seconds. The hose cart weighed 1,325 pounds.

2-17
Volunteer firemen battle the flames in this lithograph from 1895. Somewhat idealized, it depicts equipment from an earlier era.

2-17

2-16

2-16
An 1819 membership certificate from a volunteer fire company, Supply Engine No. 1 of New York City. Note the leather fire buckets being carried on poles at bottom right.

While racing to the scene and at the fire, the crews worked to a cadence shouted by their foreman. A good example is the cadence of Jefferson Engine Co. No. 6 of New York City, known as the Blue Boys: "True blue never fades." Usually the foreman stood on the engine between the brakes, giving the cadence and urging his men on. When a fireman tired and had to drop off the brakes, another volunteer jumped in to take his place. This too had to done with skill, especially when the pace was fast, or serious injury could result. Someone who jumped in awkwardly could be caught by the rising pump handle and suffer a smashed finger or arm; dislocations and fractures were not uncommon. A large, well-trained volunteer company could keep their engine in operation for hours on end.

THE LIFE OF A VOLUNTEER

The spirit behind the volunteer companies is summed up by the motto of the first fire company organized in New York, Hudson Engine No. 1 (founded in 1731): "Where duty calls, there you will find us." To be a volunteer was much more than a duty, however. Service to the community remained the highest goal of the volunteer companies, but other reasons, such as prestige, excitement, camaraderie and even the desire for political power motivated the volunteers to join. More practical considerations, such as exemption from jury and militia duty, were the motivation of some.

2-18
Sheet music covers are an important illustrative source from the days before photography. This cover is for the "Diligent Hose Company Quick Step," published in 1849. The hose wagon it depicts is decked out in parade form, with a cut-glass covering over the hose reel. Note the short capes and round hats of the firemen.

The volunteers took enormous pride in their equipment, and lavished considerable care and money on it. No piece of apparatus was complete without elaborate decoration. The leader jacket or hose cover, the box of the engine, and panels on the pump housing were often beautifully decorated or painted with scenes of special significance to the company. Artists with a flair for engine painting were sought after and paid high prices. Patriotic and mythological scenes were common, as were depictions of firemen in action. Other favored subjects came from the popular scene. In New York City, the full gamut of subject matter could be seen. Engine Co. No. 13 had a painting of Zeus hurling thunderbolts on the box; Engine Co. No. 32 had a painting of the Bunker Hill monument. Some engine decorations were presentation panels given to the company. A good example of this is a panel presented to the Franklin Fire Company of Philadelphia by the New York Fire Department in gratitude for their help during the great New York fire of 1835. Another very common subject would be a scene selected from Lord Byron's long poem "Mazeppa," first published in 1819. The poem relates the heroic adventures of passionate Jan Mazeppa, a Cossack chief, and was considered somewhat racy in its day. With liberal changes and interpolations it was made into a popular stage play. A famed actress of the period, Adah Isaacs Menken, starred in one version that had a long and successful run. Her popularity in the role of Mazeppa is probably the inspiration behind the many depictions of the character on fire engines. Perhaps the most famous (or infamous) engine painting was the head of a snarling Bengal tiger on the box of Americus Engine Co. No. 6 (built by John Agnew in 1842). By the 1840s, New York's volunteers had become a potent political force whose votes could spell success for a politician. William Marcy Tweed joined the Americus company, and in 1850 was elected its foreman. Tweed used the fireman vote to get elected to office and to build the strength of his corrupt political machine, known as Tammany Hall. "Boss" Tweed retained his membership in the Americus and kept the votes of the firemen. When cartoonist Thomas Nast began his campaign to oust Tweed and his cronies and clean up city government, he used the Americus tiger as a symbol for Tammany Hall. Nast's campaign galvanized public outrage, and Boss Tweed ended his days in jail.

The hose carts of volunteer days were naturally elegant, with graceful lines that cried out for ornate decoration. Some of the most elaborate carts were plated with gold and silver, and ornamented with jewels and paintings. Spring-mounted bells of solid silver were not unknown, and the lamps were equally fancy.

In addition to the paintings and beautiful striping, carving, and other decoration painted on the engines and hose carts, the metal work was often plated with nickel, silver or even gold. The lamps were also plated and equipped with elaborate ornamentation; the etching on the glass could rival the paintings. Kenneth Dunshee describes, in *As You Pass By*, the beautiful hose carriage of Rutgers Hose Co. No. 26, built by Van Ness: "The running gear was red, with gilt stripes. The reel was red, ornamented with beautiful gilt carving of intertwining olive and oak branches. Over the arch on each side of the reel was a miniature equestrian statue of Washington. On the front box was a representation of the Rutgers mansion and the motto of the company: 'The Noblest Motive is the Public Good.' On the back tool box was a painted portrait of Colonel Rutgers; on the side panels were small sketches of a girl coyly peeping through a lattice."

The expense of all this decoration was high—and it was paid entirely from the volunteers' own pockets. Decoration became one element of the rivalry between companies. Although the engine panels and some other decorated parts would be removed before a fire, other decoration was integral to the engine or hose reel. A washing by a rival engine could ruin not only a company's reputation but its art work as well.

The life of a volunteer fireman was exhilarating, satisfying, and often dangerous. Breathing apparatus was unknown, communications were through speaking trumpets, and protective gear was primitive.

The ideal of the perfect fireman—strong, brave, and dedicated—was already well established when the gaslights at the Park Theatre in New York City started a fire during a performance starring the famed actress Mrs. Dyott. Handsome, courageous Malachi Fallon dashed boldly into the flames and emerged to the cheers of the crowd carrying the beautiful actress in his arms. The incident, widely reported and colorfully embellished, set the standard for all firemen to come.

One of the most famous of the New York City volunteers was Zophar Mills, the foreman of Eagle Engine Co. No. 13, better known as Lucky Thirteen. Mills was the epitome of what a volunteer should, as one of his contemporaries wrote: "Where the smoke was the thickest and the fire hottest, there he was. I don't believe there has been a fire in 45 years that he has not been to." Mills had a number of near-miraculous escapes. At a fire in 1834, he was standing by a masonry wall when it collapsed. As Mills later recalled, "I suddenly saw one of the high gable walls spread out like a blanket, and coming down upon us. My only chance was to turn my back and take it; there was no time to run. I was knocked flat, of course, by the falling mass of brick, and was forced through the second-story floor, and also through the first-story floor, into the cellar. I remember raising myself on my elbows, and then getting up and walking out, after having gone through two floors with that wall on top of me."

Two firemen standing next to Mills were killed. Mills credited his lucky escape to his tin speaking trumpet, which was slung across his back; he claimed his back was black and blue for six months afterward. Not more than a year later, Mills was again carried by falling debris through two stories of a building, again ending up in a cellar, and walking out unassisted, with only the loss of his hat. New York City's first modern fireboat, commissioned in 1883, was named for Mills, and served until 1934.

2-21
Leather fire buckets remained a standard household item well into the 19th century. The bucket on the left was made in 1847 for W.H. Lovett, a member of the Beverly Union Fire Society. The bucket on the right features an allegorical figure of Hope.

"THE RACE"
[After a lithograph published by Currier & Ives, 1854]

2-19

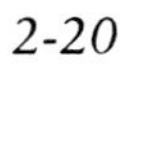

2-20

2-29

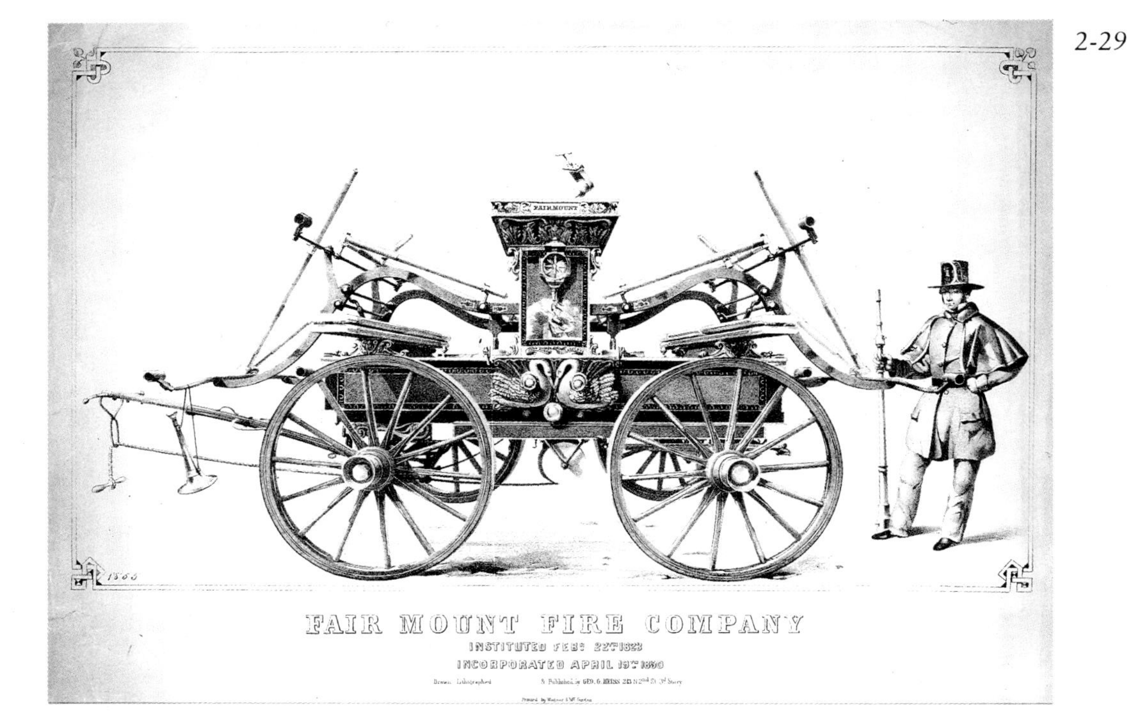

2-30

2-32

Another old New York fireman recalled a memorable escape in 1842: "A stove store caught fire on the third floor. I had two streams on the second floor, and was throwing them up the hatchway. The floor above suddenly gave way, and the weight of the stove on it carried away every staircase down to the cellar. We looked around for a means of escape, and found an old sign-board covered with a preparation of smalts—small pieces of broken glass. We put one end of the board on a window-sill, and the other down on a small outhouse, forming a very steep inclination. Then we slid down. Each of us lost the seat of his trousers, and I parted with some flesh besides, so that I didn't sit down for some time afterward."

At a fire in 1845, several firemen were on the roof when the building exploded, sending the roof and its passengers into the air. The roof sailed down to the pavement, and the firemen walked away unharmed.

When Barnum's Museum caught fire in 1865, the firemen had to cope with the fierce menagerie animals trapped by the flames. An unassuming fireman named Johnny Denham, a member of Fulton Hose Co. No. 15 (Victory-Peterson), found himself on the street faced with a Bengal tiger that had escaped from its cage. Denham seized an axe and dispatched the tiger, to the cheers of the crowd. Galvanized by this, he rushed into the burning building and carried out, singlehandedly, the 400-pound Fat Lady. He entered the building a third time and rescued an albino.

Being a volunteer fireman could entail considerable expense. Zophar Mills estimated that in one seven-year period he had spent $3,000—a very considerable sum in the first half of the 19th century. The volunteers paid for their equipment, fire houses, turnout gear, and all other expenses almost entirely from their own pockets. Crack companies could count on additional income from bonuses paid by insurance companies to the first engine on the scene. Another source of income was the fines levied on members for various offenses. In 1843, Engine Co. No. 1 of Chicago assessed members who missed a meeting 25 cents; drinking, smoking, and bad language also carried fines. Being late to an alarm was a fine of 50 cents. Absence from a fire (unless ill; being out of town didn't count) carried the heavy penalty of $2. A good enough excuse could sometimes persuade a foreman to let off an offending fireman. The minute books of the old volunteer companies contain amusing instances of firemen excused because the alarm caught them in a compromising position. On the other hand, Zophar Mills only with difficulty restrained himself from going to a fire on his wedding night, and actually did go to one early the next morning. The experience of one volunteer was not at all atypical: "I never lost a day in my business. Often I was out with my engine four nights in the week, yet I was at work as usual in the morning."

The high-spirited volunteers of any city in the first half of the 19th century developed friendly (and not-so-friendly) rivalries among themselves. At first the natural competitiveness between companies was good-natured, if sometimes foolish. For example, rival companies would sometimes race each other home

2-19
Friendly and not-so-friendly competition between fire companies was the norm during volunteer days. This engraving, titled "The Race," is based on a Currier & Ives lithograph from 1854.

2-20
Charles T. Holloway was well-known builder of fire engines in Baltimore. The "Zephyr Quick Step" was dedicated to him in the 1850s in his role as president of the Pioneer Hook and Ladder Company No. 1 of Baltimore.

2-29
The Fairmount Fire Company was instituted in 1823 and incorporated in 1850. This print of its Philadelphia-style end-stroke engine dates from 1855. Note the elaborate decoration on the engine and the nozzle held by the fireman at right.

2-30
Volunteer Fire Company No. 4, in an oil painting by an unknown artist, dating from around 1850. Slogans such as "Staunch and True" were very common among the volunteer companies.

2-32
Decorated glass panels were sometimes used to adorn hose reels. This hose reel, made in 1837, was used by the Hope Hose Company of Philadelphia. Needless to say, the glass panels were carefully stored and placed on the hose reel only for dress occasions.

from the fire. In one instance in Chicago in the 1840s, the Red Jackets and the Fire Kings raced so recklessly that both seriously damaged their engines.

By the late 1840s competition was starting to get out of hand. The old breed, who volunteered from civic pride, was giving way to a younger, tougher group who often saw the firehouse as a combination political clubhouse and saloon. Many of the new breed were just as brave as the old, but considerably rowdier and less mindful of responsibility. The companies could respond only feebly to accusations that their members included drunkards and hoodlums.

The competition for water with which to fight the fire brought out the worst in the firemen. Water was often the weak link in the firefighting chain. In the days before municipal water supplies, wells, cisterns, ponds, and rivers were the sources used, and these were not always at hand or sufficient. Indeed, during the great New York fire of 1835, one building was saved when the resourceful firemen pumped vinegar from barrels in a nearby warehouse onto the flames. The engine that arrived first laid claim to the water, and was willing to fight to hold onto it. Later, when wooden water pipes were laid under the streets, the first firemen on the scene would dig down to the pipe and drill a hole in it. The engine's suction hose would then be inserted. When the fire was out, the firemen plugged the hole with a piece of wood and marked the spot. This gave rise to two lasting bits of fire lore: the word fireplug as a synonym for fire hydrant, and the pointed end of a fireman's axe, used to remove the wooden plugs.

Further complicating the issue was the alarm system. When a fire was detected, bells in the watch towers were rung. The number of tolls indicated the location, not always with accuracy. The nearest fire company was generally first on the scene, but other companies would also converge on the fire, often unnecessarily. Brawls over precedence and hydrants would break out, and sometimes the companies would fight while the building burned down. Usually, however, the firemen would extinguish the flames and *then* have the brawl. One old fireman reminisced fondly about a famous brawl, remarking "Pipes, axes, and any weapon on hand was used in the fight. It was a terrific fight and lasted a long while." A number of fire companies were "turned tongue in," or disbanded, as punishment for fighting.

By the late 1840s the situation in many cities was getting out of hand. The firemen were as much a menace to the public as a help, and they were stubbornly resistant to new developments in equipment. One clear solution presented itself to city fathers throughout the country: the time for a paid, professional fire department had arrived.

2-31

2-31
This well-preserved hose reel was used by the United States Company of Philadelphia. It was built by D.G. Matthews & Co in 1849.

THE AGE OF STEAM

The use of steam engines to convert heat energy into mechanical energy was first proposed by Hero of Alexandria. Little is known of Hero, who wrote in Greek and was active around A.D. 62. As a mathematician, Hero made many contributions to the measurement of geometric figures. As an inventor, he developed mechanisms that operated using compressed air, water, and steam, including a primitive sort of steam fire engine. The true development of steam as a conversion medium for engines languished for centuries, however, to be rediscovered and advanced only toward the end of the 17th century. In England in 1698, Thomas Savery patented a steam-powered water pump that didn't work particularly well. Around 1711, Thomas Newcomen developed an atmospheric steam-powered engine that was also used to pump water. The basic principle behind Newcomen's work was the use of steam to move a piston back and forth within a cylinder. The back-and-forth motion of the piston can be converted into rotary motion for driving machinery through the use of a crankshaft. The Newcomen steam engines were inefficient, however, and were no serious challange to muscle power, whether human or animal.

While making repairs to a Newcomen engine, the great Scottish inventor James Watt made several seminal improvements that made steam engines a practical reality. He combined the improvements into a new type of engine patented in 1769. The new steam engine featured a separate condensing chamber for the steam, an air pump to bring the steam into the condensing chamber, and insulated engine parts. Watt also used steam pressure to move the piston in both directions within its cylinder, providing marked improvements in efficiency.

Newcomen steam engines were originally developed to pump water from mines. Watt's steam engines were very quickly put to the same use, but it would be decades before anyone thought of using steam-powered pumps to fight fires.

By the end of the 18th century, many experiments in steam-powered ships were taking place. The first true steamship was developed by the American inventor and engineer Robert Fulton. In 1807 his ship *Clermont* gave a convinc-

3-1
A Waterous No. 1 size steam engine from 1886. This apparatus weighed 2,200 pounds and could pump 250 gallons per minute.

3-2
The Waterous No. 3 size steam engine from 1886 weighed 2,800 pounds and was meant to be drawn by horses. It was also available fitted out with a hand pole and drag rope. Standard equipment included a 10-inch brass gong, a wetting-down hose, and a thawing-out hose.

3-3
The No. 4 and No. 5 size steam engines from Waterous had capacities of 400 and 600 gallons per minute respectively. These models from 1886 were meant for use by paid departments.

ing demonstration of steam's practicality when it made the 150-mile trip from New York City to Albany in the amazingly short time of just 32 hours. By 1819, the first steam-powered ships had crossed the Atlantic.

Dependable steam locomotives for railroads were being manufactured in America by the 1830s, and by 1835 there were 175 steam locomotives in service in the United States. The use of steam to power pumps and other machinery was well established. Given the widespread acceptance of steam machinery for industry and transportation, the development of steam-powered fire engines came oddly late. The regular use of steam fire engines came even later, over the strenuous objections of the volunteer firefighters.

THE EARLIEST STEAM FIRE ENGINES

The first true steam fire engine was developed in London in 1829 by a hydraulic engineer named John Braithwaite, assisted by John Ericsson. Born in 1797, Braithwaite was an expert in the manufacture of high-pressure steam engines. John Ericsson was born in Sweden in 1803 and trained as a military engineer. He went to London in 1826, where he worked with Braithwaite on various projects utilizing steam power. (Ericsson is best known, however, for his later invention of the screw propeller and for designing the Civil War battleship *Monitor.*) Together Braithwaite and Ericsson designed the first railroad locomotive to travel a mile in under a minute (56 seconds in 1829). They also designed an effective steam fire engine.

The Braithwaite apparatus was heavy and clumsy, but powerful. The horizontal steam cylinders had a bore of 7 1/2 inches, while the pump cylinders had a bore of 6 1/2 inches. A single piston rod operated both cylinders, with the pump cylinder at one end and the steam cylinder at the other. (This pioneering design element would later be incorporated into most later steam fire engines.) The boiler, fueled by coke, was at the rear; it took about 20 minutes to get up steam. The pump was located at the front, just behind the driver's seat, and topped with a large, spherical air chamber. The suction and discharge hoses were attached just beneath the pump. The pump could discharge an impressive 200 to 250 gallons per minute at a pressure of about ten pounds per square inch (psi), and generated ten horsepower. By comparison, the largest hand pumpers of the period, manned by a full complement and stroking at the normal 40 to 60 strokes a minute, could discharge only about 150 gallons per minute. Working at the top speed of 120 strokes a minute, a large hand pumper could discharge considerably more, but only at the cost of quickly exhausting the firemen. Even with numerous substitutes to jump in as the firemen fell out, a hand pumper could not continue long at top speed.

Braithwaite hoped to have his new steam engine, christened *Novelty*, accepted by the underwriters of the British insurance companies. At a serious London fire in the winter of 1830, his engine continued to pump for five

Nos. 1 and 2 Engine.

3-1

No. 3 Engine.

3-2

Nos. 4 and 5 Sizes Engines (Full View).

3-3

3-42

3-43

straight hours, long after the hand engines at the scene had frozen up. The engine was used at another serious fire in 1830, and was used again when the Houses of Parliament burned in 1834. Inexplicably, the underwriters were not impressed. Braithwaite's hopes in England were dashed by the rejection not only of the underwriters but of the public; his first engine was destroyed by a mob (the British did not begin using steam fire engines until 1858). He did build several more steam engines, however. A smaller, single-cylinder engine generating only five horsepower was contsructed in 1831. An engine that went to Liverpool was built around 1831; this apparatus had a two-cylinder horizontal engine, but experimented with gearing between the piston and the pump. Another Braithwaite engine was delivered to Berlin for the King of Prussia in 1832. The Berlin engine had two cylinders and two pumps, and could supply up to four hose lines. It generated 15 horsepower, weighed four tons, and could pump about 300 gallons in a minute. Braithwaite built a fifth engine in 1833, but gave up on fire engines after that and concentrated on locomotives.

In America, serious fires with significant losses were occurring with increasing frequency in many cities. In 1839, insurance losses in New York City were $720,000; in 1840, they had skyrocketed to well over $4.5 million. Better fire protection from more powerful fire engines was obviously needed. The insurance companies grouped together and commissioned an engineer named Paul Rapsey Hodge to build a steam fire engine.

Hodge had come to America from England sometime before 1837, perhaps bringing with him some knowledge of Braithwaite's designs. He worked as a draftsman for Thomas Rogers, an early locomotive builder. Legend has it that Hodge was fired for incompetence after the firebox he designed would not fit into the boiler shell.

Whatever his faults as a draftsman, Hodge went on to become a well-known expert on steam engines. He was the author of *The Steam Engine*, published in New York in 1840, which was the first book to contain drawings of an American locomotive. Hodge demonstrated his fire apparatus on 27 March 1841, in front of City Hall. Called the *Exterminator*, the self-propelled engine was 13 1/2 feet long and weighed over 14,000 pounds. The large wrought-iron drive wheels at the rear were driven by piston rods connected to the double-acting horizontal pumps, which were bolted directly onto the cylinders. With its large horizontal boiler and steam cylinders, separate smokestack and smaller front wheels for steering, the whole apparatus bore a strong resemblance to an early steam locomotive. More particularly, it bore a strong resemblance to locomotives built by H.R. Dunham & Co., for whom Hodge had worked. The Dunham locomotives, in turn, had a strong resemblance to the better-known Baldwin locomotives of the period. (It is possible that Hodge also worked for Baldwin at some point.) The boilers and weight (about seven or eight tons) of Hodge's steam fire engine were very similar to those of a Dunham locomotive.

In operation Hodge's engine was as clumsy as it looked. It took half an hour

3-42
This 1890 Wilcox hose reel was meant to be hand-drawn. Note the wooden tool box attached to the rear. Modern fire helmets cover two nozzles; a third is stored along the shaft.

3-43
A well-preserved hose tender at a muster in Valhalla, New York. Note the gong under the driver's footboard.

3-4
Side views of the Waterous No. 4 and No. 5 engines. The hump in the frame is designed to allow the front wheels to turn under the body, making for a smaller turning radius and easier maneuverability. Standard equipment with these engines included 20 feet of best wire-lined smooth-bore suction hose, copper strainer, large fuel tender of ample capacity with hand rails, driver's seat with cushion, one 12-inch gong with foot-trip attachment, foot brake, two brass lanterns, brass torch for engineer's use, oil cans, poker, fire shovel, and a full assortment of wrenches.

to get up steam, and the rear wheels needed to be jacked up to act as flywheels. On the other hand, at its first public trial the engine shot a stream 166 feet high.

No engine company wanted to accept the newfangled machine. Finally, the men of Pearl Hose Co. No. 28 in the Fifth District were persuaded to try it out. Unenthusiastic from the start, they quickly developed a long list of complaints about the engine. Some were valid: In addition to the time it took to fire it up, the machine was heavy and broke down often (in part perhaps because the men assigned to its maintenance didn't understand it). Some were exaggerated: The men feared the boiler would explode, and sparks from the smokestack could cause fires. More persuasive, perhaps, was that the other fire companies refused to have anything to do with the Pearl's new machine, and would not supply water to it.

The trial lasted just a few months. The new engine was thankfully returned to the insurance companies, who it turn sold it to a box manufacturer for use as a stationary engine. Discouraged perhaps by the failure of his fire engine, Paul Rapsey Hodge returned to England in 1847. He published another book, *Analytical Principles and Practical Applications of the Expansive Steam Engine* in 1849, and died in England some time after 1869.

STEAM TAKES OVER

After the abortive experiment with Hodge's engine, the development of steam fire engines languished. The volunteers around the country continued much as before, using hand engines and growing increasingly rowdy. Finally, an incident in Cincinnati changed the face of American firefighting.

A nasty fire broke out in a mill in Cincinnati one night in the fall of 1851. The battle against the flames began with Washington Hose Co. No. 1 and Western Fire and Hose Co. No. 3; eventually ten other companies gave assistance. Another volunteer company, located across the Ohio River in Kentucky, saw the flames and crossed the river, ostensibly to help its sister company, the Washingtons. The situation quickly degenerated. The mill was destroyed when 13 companies stopped fighting the fires and engaged instead in a general brawl.

While the Cincinnati example is extreme, it was by no means unique to the time. In many places, the volunteers had changed from being dutiful citizens to being little more than gangs who hung around the firehouse and voted as a bloc. They were a cocky group, secure in the knowledge that no matter how much they drank and fought, they also put out fires for free—which made them indispensable. This arrogant attitude was in for a rude shock.

The key to disbanding the volunteers and replacing them with a paid fire department was finding a way to fight fires effectively with fewer men. After the disgraceful actions of the volunteers in Cincinnati, the city fathers, already familiar with steam power because the city was center for steamboat and locomotive construction, decided to explore seriously the possibilities of steam fire engines.

3-5
The Twin Falls, Idaho, Fire Department poses in the late 1880s. Engine Company No. 1 is at left; a combination ladder and chemical wagon is at right.

3-6
The Newark, New Jersey, Fire Department in action in 1905. Two things are unusual about the horses in this picture: one is a different color, and the horse on the left is out of step. For esthetic reasons, firemen preferred the horses to be all the same color. For practical reasons of smoothness and control, the horses were trained to move in step with each other.

3-7
This 1908 Ahrens steamer once served the Detroit Fire Department. It has been restored by the Box 42 Associates of Detroit and is a popular parade piece. It is shown here at a muster at the Henry Ford Museum in Dearborn, Michigan.

3-8
The very handsome steam engine of Laurel Engine Company No. 1 in Pennsylvania is on display in this 1911 photo. Note the interesting stag emblem mounted on the air chamber dome.

3-9
Union Steam Fire Engine Company No. 3 poses outside headquarters in 1911.

3-10
NYFD Engine Co. No. 205 makes a pre-arranged run on 20 December 1922. The horses are dashing through downtown Brooklyn toward Borough Hall, where a new gasoline pumper awaits them. This would be the last horse-drawn run in New York City.

3-5

3-6

3-7

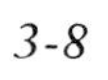

3-8

3-9

3-10

3-44
A chief's buggy from 1809, still on the parade circuit.

Two entrepreneurs in Cincinnati had come up with an idea for a fire engine with a steam-powered pump. The first was Alexander Bonner Latta, an expert machinist and inventor who had built and operated the first railroad locomotive west of the Alleghenies in 1845. The second was his partner in the venture, Abel Shawk, a mechanic and metal worker. Shawk incorporated a new copper coil design into the engine's boiler; this raised steam quickly. Latta supplied the boiler and his expertise with steam propulsion. The apparatus they came up with was cumbersome, but it worked. At its first public demonstration, the boiler raised steam in under five minutes, and the pump threw a stream 130 feet through 350 feet of hose. The city council was convinced. Led by Councilman Joseph Ross, they appropriated $5,000 to purchase a new steam fire engine from Latta & Shawk.

The volunteers were not happy to hear this news, and they were made even more unhappy by the information that legislation to create a paid department was pending in the Cincinnati city council. Latta and Shawk worked in secrecy and feared for their lives, while the volunteers demonstrated (or rioted) in opposition.

The new engine incorporated several interesting new design features, many at the suggestion of Miles Greenwood, a very experienced engineer and volunteer firefighter. The new machine, named *Joseph Ross* after its sponsor (and later nicknamed *Uncle Joe Ross*), was self-propelled, more or less. It weighed in at 22,000 pounds, and was so heavy that four horses were needed to pull it

3-45
An American LaFrance steam engine from 1890. The heavy draft horses pulling this rig at a muster in Maryland are handsome but inauthentic: horses this ponderous would never have been used in actual fire service.

while it raised steam; even after the self-propulsion gear was engaged, the horses were still needed, especially when going uphill. The basic design was still that of a steam locomotive. The steam cylinders were coupled to the two rear wheels, driving them in the same way as a locomotive. Steering was provided by a single wheel in the front. The same steam cylinders, when the rear drive was disengaged, were used to drive the double pumps.

Latta and Shawk demonstrated their new engine on 1 January 1853, in a competition with the most powerful hand engine in the city, the Union Fire Company's Hunneman. The contest began when the steamer arrived at the site, nearly 20 minutes after it left its engine house; it could be heard rumbling over the cobblestones long before it was seen. While the boiler built up steam, the Union volunteers set up their hand engine and began pumping, demonstrating how quickly they could get into action. Their best efforts produced a 200-foot stream. When the Uncle Joe was finally in action, it produced a powerful 225-foot stream. And while the volunteers gave up in exhaustion half an hour later without bettering their stream distance, the steamer kept pumping. Even worse, from the volunteers' point of view, the steamer kept pumping with undiminished strength even when six streams were working.

The end had come for the volunteers. Over loud objections, on 10 March 1853 the Cincinnati city council voted to have paid firemen—the first salaried fire department in America—starting 1 April of that year. Miles Greenwood was appointed the first chief.

A year later, a group of public-spirited citizens raised the money to purchase another steamer. By now Shawk and Latta had dissolved their partnership over differences in design philosophy. Latta alone built this engine. Named the *Citizens' Gift*, it cost $13,400. This apparatus too was ostensibly self-propelled, but still needed a complement of four horses to move it around. It used a double-cylinder, double pump system. A contemporary account of an exhibition performance states that "she ran four squares [blocks] and was throwing two fine streams through 200 feet of hose that had been laid and run up a ladder five stories high, in precisely six minutes from the tap of the bell, at which moment she was standing cold and unhitched in the engine-house. Her best throw was a horizontal stream from an inch and a half nozzle a distance of 287 feet. To work this machine, including the horsemen, requires the services of ten men."

That same year, one of the volunteers' fears about steam engines did come to pass: the boiler of the *Uncle Joe Ross* exploded, killing the engineer.

By 1858, Cincinnati had three large steam engines and one small one in service, and the paid department was fully manned. The department consisted of a one chief engineer, two assistant engineers, and one captain and two lieutenants for each 30-man company. Before the steamers came into service, the 28 hand engines in Cincinnati could pump a combined total of 121,200 gallons an hour, using the work of several hundred men. The three large steamers combined, by contrast, could pump 178,500 gallons in an hour, and required only 30 men at most.

Many other cities were quick to appreciate the virtues of steam. A Latta steamer named for Miles Greenwood was demonstrated in both New York and Philadelphia in 1855, but ended up being purchased by Boston. By 1861 the entire Boston department had converted to steam.

In February 1855 another Latta steamer was tested in New York City against the famous side-stroke *Haywagon* of Empire Engine Co. No. 42. It took the Latta engine eight minutes to get up steam. The *Haywagon* pumped a horizontal stream 189 feet; the steam engine managed 182 feet. The *Haywagon* also managed to pump a higher stream, but the men were afterward completely exhausted. The steamer, of course, was able to continue for hours. Zophar Mills, the dean of New York City firemen, was deeply impressed. New York City finally purchased two steamers (at a cost of $17,000 each) in 1859, but these never went into regular service. A steamer known as the *Elephant* was donated to the city by the fire insurance companies in 1859 and assigned to Manhattan Engine Co. No. 8. It thus became the first steam-powered fire engine in regular service in New York City. This machine, built by Lee & Larned of New York City, had a rotary pump and weighed in at 5,600 pounds, much lighter than earlier steam engines because brass and steel were substituted for many iron parts.

St. Louis went to steam engines and a paid department in 1857. Brooklyn and Rochester had steam engines by 1858, and Milwaukee was using them in

3-11
Steam engines were still something of a novelty in 1872, when this illustration appeared in The Graphic.

3-12
This photo of Laurel No. 1 of York, Pennsylvania, was taken in 1911. Note the preconnected suction hose wrapped around the rear.

3-13
Royal Hose and Nozzle Company No. 6 shows off its turret gun and other gear in front of the fire house in 1911. A Dalmatian dog is seated next to the driver.

3-14
Fire Department headquarters in Seattle, Washington, in 1890. A fire tower with a huge bell is attached to the building, and a tobacco store is conveniently located next door.

3-15
The engineer studies his gauges at a fire on Broadway in New York City in 1893.

3-16
An Amoskeag steamer being demonstrated during the construction of the World's Columbian Exposition in 1891.

3-11

3-12

3-13

3-14

3-15

3-16

3-46

WASHINGTON
ENGINE CO. 2
RIVERHEAD F.D.
NEW YORK

FREEPORT
EXEMPT
FIRE DEPT.
AMERICAN LaFRANCE

3-47

1861. Paid departments followed soon after. In February 1858, Chicago purchased a steam engine, called *Long John* after Mayor John Wentworth. On 6 March, some volunteer companies organized a disorderly protest against the steamer in the courthouse square. The firemen were quickly dispersed by the police and their equipment was confiscated. On 22 March, the offending companies were disbanded, and in August Chicago went to a paid department. The pay for a fireman was $25 a month.

In May 1858, the members of the First Baltimore Hose Company purchased a steam engine from Captain John Ericsson, who had docked in Baltimore while transporting the engine to Norfolk.

3-55
A close-up look at the flywheel from a 1904 American-LaFrance steam engine.

In Phildelphia, a Latta steamer was presented to the city by a group of citizens in 1855. This behemoth did little to aid the cause of steam: it weighed 20,000 pounds, needed the services and expense of horses to pull it and two salaried engineers to run it, and ran up $20,000 in maintenance and repair bills within three years. Even so, in 1857 the Philadelphia Hose Company had purchased a steam engine; the Hope Company got one in 1858; and the Hibernia Company, one of the city's oldest and proudest, had a steamer by 1859. Other companies followed suit, and by the end of 1859 the city boasted 20 steamers, all operated by volunteers. Opposition to steam engines ran high in New Orleans. The underwriters for the city purchased an early Latta steamer in 1855; in turn, the city council voted for a paid department that same year. The Latta machine was known as *Young America*, and weighed in at 18,000 pounds. It was an unwieldy monster that never performed up to expectations, but nonetheless by 1861 six New Orleans companies were running steam engines.

In New York City, the politically powerful volunteers stubbornly resisted change. Because many politicians in the city owed their positions to the volunteers, the switch to a paid department was a long time coming. As Chief Greenwood of Cincinnati once said, "Steamers never get drunk. They never throw brickbats. Their only drawback is that they can't vote." After a drawn-out battle through the state legislature and the courts, however, on 31 July 1865 the volunteers gave way to a paid Metropolitan Fire Department. When the volunteer companies turned over their equipment to the professionals, the tally included 32 hand-drawn steam fire engines, and one horse-drawn steamer. The new department consisted of about 500 men under the leadership of Chief Engineer Elisha Kingsland. The annual salary for an ordinary fireman was $700 a year.

In Philadelphia, the birthplace of the volunteer and the home of fire engine innovation, the transition to a paid department took even longer. The city's volunteers were replaced by professionals in 1871.

3-46
This American LaFrance third-class Metropolitan steam engine was delivered to the Riverhead, New York Fire Department in 1903.

3-47
Steam fire engines were solidly built and never wore out. Rather than junk perfectly good apparatus, fire chiefs substituted tractors for horses. Here a 1906 American LaFrance steam engine is hitched to a 1916 American LaFrance tractor. This rig is owned by the Freeport, New York Exempt Fire Department.

THE SELF-PROPELLING STEAM FIRE-ENGINE, "J. C. CARY."

3-17

3-18

THE PRIDE OF HARTFORD

3-19

TWO-WHEEL ENGINE TENDER.

Built by the AMOSKEAG MANUFACTURING Co., Manchester, N. H. Represents those used in the Metropolitan Fire Department by Engine Cos. Nos. 2, 7, 8, 11, 16, 19, 22,

3-20

AMOSKEAG MANUFACTURING CO. MANCHESTER, N. H.

3-21

First Steam Fire Engine of Carlisle, Pa., owned by the Hose Company.

3-22

THE STEAM APPARATUS MAKERS

The second steam engine built by Abel Shawk and Alexander Latta in 1852 ended up costing $10,000, double what the Cincinnati city council had appropriated. The two men apparently argued about the design and expense of their new engine, and decided to go their separate ways when it was completed. By 1855 Abel Shawk had set up on his own as a manufacturer in Cincinnati, calling his firm the Young America Works. The first fire engine he built was called the *Missouri.* Shawk had given up on self-propulsion, for this engine was horse-drawn. Little is known of his later engines or the fate of his company.

Alexander Latta and his brothers formed a new company called A.B. & E. Latta. Alexander continued to design boilers, railroad locomotives, and steam engines of various sorts, and received several patents on important improvements. He remains best known for his work with steam fire engines. In 1855 he received a patent for a self-propelled, three-wheeled apparatus. The rear wheels were connected by piston rods to the same steam cylinders that operated the pumps. When the engine arrived at the fire ground, it was jacked up and the rear wheels became the flywheels. By 1860, the firm had built some 30 steam engines, including engines for Memphis and Nashville. Alexander Latta retired from business in 1862, and died in 1865.

Following the success of Latta, many makers of hand engines turned to steam. The technology and manufacture of steam engines, however, was considerably different—and much more complex—than that of hand pumpers. Questions of boiler size, cylinder design, pump placement and many more questions needed to be answered through careful design and practical experience. Poor design was not only inefficient but dangerous, since the boiler could explode with great force. Unlike hand engines, which used a great deal of wood in their construction, steam engines were built almost entirely of metal. For the builders of hand engines, the transition to steam manufacture involved a considerable capital outlay for new design and manufacturing equipment, as well as the necessity to hire mechanics and machinists skilled in the area. In modern terms, a comparable transition would be from making vacuum tubes to making microchips. Not surprisingly, many of the best-known makers of hand engines failed when they turned to steam engines.

John Agnew of Philadelphia built only four steam engines. One, the *J.D. Danbury* built in 1860 for Memphis, Tennessee, was an unusual but clumsy design that utilized a horizontal boiler and two smokestacks. Another, the Fairmount No. 32, was built for Philadelphia and utilized a more conventional design with a vertical boiler.

Agnew quickly found himself in competition with several new manufacturers in Philadelphia. Among them was a firm called Reanie & Neafie. The firm's first steam engine was built in 1857 for Philadelphia. Christened the Philadelphia No. 1, this engine weighed in at nearly 7,500 pounds. It had a vertical boiler jacketed in wood and bound by brass bands, and a tall stack; the

3-17
This monstrous self-propelling steam was christened the J.C. Cary *and demonstrated in New York City in 1858.*

3-18
A.B. & E. Latta advertised their steam fire engines in this ad from the 1850s. Note that this was three-wheeled rig; the single front wheel was used for steering.

3-19
Hartford's famous Jumbo, *the largest steam fire engine in the world, in a photo taken in the 1890s. This huge machine, built by Amoskeag in 1889, was self-propelled and weighed over five tons.*

3-20
A two-wheel engine tender built by Amoskeag for the New York City Metropolitan Fire Department in the mid-1860s. This tender carried a hose reel and a bin for extra coal.

3-21
The complete Amoskeag line of firefighting equipment, circa 1872. Although steam engines predominate, the company also advertises its engine tenders and hose reels.

3-22
Sadly neglected in this photo from 1909, the first steam fire engine of the Carlisle, Pennsylvania fire department languishes in a back lot.

3-57
The engineer's view of a 1904 Nott steam fire engine. The wooden lid over the coal box would be lifted away when the engine reached its destination.

front-mounted pump had an unusually tall air chamber. This pumper remained in service for more than 40 years. The firm of Reanie and Neafie made 33 fire engines before dissolving in 1870. (At some time after that, Mr. Neafie went into partnership with a Mr. Levy of Philadelphia and built at least one fire engine.) Another well-known maker of hand engines, James Smith of New York City, built only 14 steamers before giving up. Several of his machines were sold to the city in the mid-1860s. Built along very conventional lines, the Smith steamers had horizontal cylinders and pumps. William Jeffers of Pawtucket, Rhode Island, another maker of hand pumpers, switched to steam engines with some success, producing over 60 before leaving the business in the 1880s. Jeffers may not have been competitive because his design was too complicated. One well-known engine produced by Jeffers in 1874 was officially named the *Julius Runge*. It soon became known as the *Jig-Saw*, however, because of its intricate system of driving and connecting rods. This fourth-class engine, built in 1874, had a cylinder diameter of 8 inches and a stroke of 7 inches; it weighed in at 3,500 pounds. The connecting/driving system involved a rod that ran from the pump piston to the steam piston; from there, the rod went up through the top of the steam cylinder and attached to a crossbar. From the crossbar a long driving rod connected to the flywheels. The venerable Button Fire Engine Works, now called Button & Blake, began producing steam engines in 1862, 30 years after its founding. From then until 1891, when it became part of the American Fire Engine Company, the Button works made more than 200 steam engines. In the 1860s the Button engines generally utilized a front-mounted horizontal pump, often with a double-domed air chamber. However, in the 1880s the firm used double pumps placed amidships below the frame for the larger engines, including the first, second, and third sizes (the first size weighed about 7,000 pounds). For the smaller fourth, fifth, and sixth sizes, the pump and engine were front-mounted. The sixth was the smallest size Button made; hand-drawn, this engine weighed only 2,500 pounds. Oddly, Button continued to use horizontally mounted piston pumps long after all the other competing manufacturers had switched over to vertically mounted pumps.

One of the most widely known hand-engine makers was Hunneman & Co. of Boston. Hunneman too began making steam engines, and produced all told about 30 from the 1860s to the mid-1880s. The firm made some handsome, sophisticated engines, but it too found it difficult to compete with the new firms springing up.

It should be remembered that through the end of the 19th century and even into the 20th, hand-pumped fire engines continued to be built and used, often by companies such as Button and Hunneman (or their successors) that had been in the business for many years. For small towns and villages with volunteer fire departments, no municipal water supplies, and limited funds, the hand pumper was cheap to buy, inexpensive to run and repair, and required little in the way of maintenance or training to operate. In remote places far from convenient and inexpensive shipping (on the frontier, in mining towns, and so on),

3-57

3-23

3-24

3-25

3-26

3-27

3-28

a local blacksmith, working perhaps with a local carpenter or carriage maker, could easily build a serviceable hand fire engine for very little money. The Eureka Hook, Ladder and Bucket Co. No. 1 of Rockville Centre, New York (then a rural village), for example, purchased a hand pumper from a local blacksmith for $150 in 1876. As nearby cities and large towns switched to steam, smaller towns were often able to purchase lightly used hand engines at attractive prices. The hand pumper also continued in use in many industrial settings, providing fire protection for factories, lumber yards, farms, and the like.

TECHNOLOGY SPURS INNOVATION

As steam engines replaced hand engines, and as professionals replaced volunteers, the firehouse changed. The new steam engines were often even more elaborately decorated than the old hand engines had been—after all, there were more parts, especially shiny metal parts, to be decorated. Steam engines were mechanically far more complex than hand engines, however, and proper maintenance was not only more involved but more important to prevent dangerous malfunctions (such as boiler explosions). Easy accessibility to key parts was an important feature stressed by salesmen.

In the early days of steam power the time needed to raise enough steam pressure to operate the pumps was often the subject of complaint. Hand engines could go into operation almost immediately after arriving at the fire ground. As boiler design advanced and the time to raise steam from a cold start dropped to just a few minutes, this criticism was sharply muted. Even so, the time lag was a problem, but one that was solved with typical fireman's ingenuity. A New York City fireman named William Gleason invented a system to keep the boilers always at the ready. Gleason's design called for a hot-water heater in the basement of the firehouse to be connected to the boilers of the idle engines. In this way hot water was constantly circulated through the boilers of the engines; a damper on the smokestack was often used to help keep the heat in. Steam could thus be raised much more quickly—in some cases in a just over a minute—when the engine was called out. There was no need to waste time uncoupling the hot-water heater and the boiler; as the engine pulled out, the coupling snapped apart and a clapper valve shut off the flow.

In large cities the engineers kept the firebox ready to light with kerosene-soaked rags or cotton. The most sophisticated fire houses had a sort of gas-fueled pilot light in the floor in front of the boiler; as the engine passed over it, the kindling was ignited.

Volunteer companies in small towns used a different system. The firebox of the steam engine while it was in the station was kept ready with coal and tinder, needing only a match to start the fire and begin heating the water. A nickel was left on the engineer's step behind the firebox. The first boy to hear the

3-23
Trailing a cloud of smoke, the Wausau, Wisconsin fire department's steam engine rushes to the fire in 1913.

3-24
The controversy over whether a horse lifts all four feet off the ground when galloping is resolved by the horse on the right in this picture from 1909. Already belching a cloud of smoke, this engine has just pulled out of the fire house.

3-25
The three beautifully matched horses pulling High Pressure Wagon No. 6 of the FDNY are in perfect step in this shot, taken in 1916.

3-26
Royal Engine Company No. 6 of York, Pennsylvania, poses for a picture in 1911. The Amoskeag steam engine dates from 1903.

3-27
The Hummingbird, *the Silsby steam pumper of Goodwill Engine Company No. 5 of York, Pennsylvania, in 1911. The rotary pump is mounted horizontally amidships on the straight frame.*

3-28
The beginning of the transition to motorized apparatus can be seen in this photo from 1910. The fire apparatus at the Central Fire Station is all horse-drawn, but the chief has an automobile.

3-56
A 1904 Nott steamer. Note the unusual frame, combining crane-neck and straight elements.

3-58
A good view of the frame construction on a 1904 Nott steamer.

alarm, run to the engine house, and light the fire got the nickel.

The introduction of horses into the station house began in the late 1860s, when steam rigs became too heavy to be hand-drawn. Horses added a completely new dimension to the life of the firefighter (who by now was probably a professional). The horses needed to be carefully trained and tended, and a deep affection grew up between the animals and the men who cared for them. Fire chiefs searched far and wide for horses perfectly matched in color and size. The ideal fire horse was at least 15 hands high (roughly 60 inches at the shoulder), and weighed at least 1,100 pounds. He or she (there was no preference) was trained chiefly on the job by the firefighters, with the help of the experienced horses. A well-trained team could be hitched to the engine in a matter of seconds, and would burst from the station house at a full gallop in perfect step with each other. Once at the fire, the horses would stand quietly in the face of smoke, fire, water and the general chaos of a fire scene.

The first horses in the fire house were kept hitched to the engines, but this system quickly gave way in the 1870s to quick-hitch harnesses that hung from the ceiling and were dropped onto the horses when the alarm was sounded. The horses could be harnessed and the engine out the door within two to three minutes.

Getting the rig out the door was speeded up by the invention of the fire pole. Previously, men got from the bunk room upstairs to the fire engines downstairs by running down stairs or, sometimes, by sliding down a chute. Captain David Kenyon of the Chicago Engine Company No. 21 came up with the idea of a sliding pole. It was put into action on 21 April 1878. The idea quickly caught on, and fire poles soon became standard equipment in fire houses.

NEW MANUFACTURERS

The Amoskeag Company originally made locomotives and steam-powered machinery for manufacturing textiles. The move into steam fire engines in 1859 was a logical—and lucrative—extension of the business. Amoskeag steamers quickly became known for their desireable qualities: They were effective, reasonably priced, and durable. In 1865, the New York City Fire Department decided to purchase 15 new steam engines from Amoskeag. New York joined Detroit, Boston, Chicago, and other cities in choosing Amoskeag. The choice, however, meant that some of the struggling manufacturers established in the 1850s gave up on making steam fire engines. By the end of the 1860s, the shakeout left Amoskeag and the Silsby Company of Seneca Falls, New York as the two major manufacturers, with Clapp & Jones of Hudson, New York, and the Button Company as secondary makers.

The first 11 engines made by the Amoskeag Company were referred to as "mongrels," because they used two reciprocating pistons to drive a rotary pump. The very first of these engines was sold to the city of Manchester, and remained in active service for 17 years. The firm then switched to a single-pis-

3-56

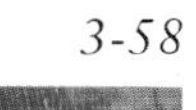

3-58

3-29

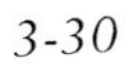

3-30

3-31

3-33

3-32

3-34

ton pump, and then quickly began offering a double-piston, double-pump model. Altogether, Amoskeag built over a thousand steam fire engines in a surprisingly wide variety of sizes and types. Crane-neck and straight frames were available; tanks could be U-shaped, harp-style, or straight. The sizes ranged from extra first to fourth.

By the mid-1860s an approximate standard for steam engine sizes had evolved, although there were minor differences among manufacturers. Generally, a double extra first had a pump capacity of 1,300 gallons per minute (gpm); an extra first pumped 1,100 gpm; a first pumped 900-950 gpm; a second pumped 700-800 gpm; a third pumped 600-700 gpm; a fourth pumped 500-600 gpm; a fifth pumped 400-500 gpm; and a sixth pumped under 400 gpm.

Amoskeag engines quickly became the standard by which others were judged. The company was a consistent innovator in the field. Among the more interesting Amoskeag developments was a series of self-propelled steamers built from 1872 to 1908. These 22 engines were massively powerful, categorized as double-extra first size; they could pump up to 1,800 gallons per minute. The first self-propellers were sold to Boston and New York City in 1872. These behemoths had a serious flaw, however: the chain drive used for transmission was located on one side of the engine, and caused serious steering difficulties, particularly when going around corners, because both rear wheels rotated at the same speed. The solution was the invention of the differential gear to transmit driving power to the rear wheels. This gives Amoskeag an interesting place in automotive history, since the firm was the first to use a differential gear on a self-propelled vehicle.

Driving an Amoskeag self-propeller was a two-man operation. The engineer stood on the fuel pan behind the boiler and operated the throttle to provide power; the driver sat in front and wrestled with the enormous steering wheel. New York City ordered four more self-propellers in 1874. These units remained in active service until 1884. At that time, two were converted to horse-drawn, and the others were placed in reserve. Other self-propellers were sold to Boston, Hartford, Detroit, Chicago, Vancouver, Pittsburgh, and elsewhere. Several remained in service until replaced by gasoline-powered vehicles. One of the best-known of these engines was the one sold to Hartford in 1889 and nicknamed, with good reason, *Jumbo*. This ponderous vehicle weighed 5 1/2 tons, and was at the time of delivery the largest fire engine in the world. *Jumbo* became something of a tourist attraction for Hartford. At full throttle it could get up to 25 miles an hour. *Jumbo* had been in service for over 30 years when it was converted to motor power by having a tractor attached to it; as such it continued in service for several more years.

In 1872 the Amoskeag Manufacturing Company was incorporated into the Manchester Locomotive Works. In 1900 the fire engine division became part of the massive consolidation that resulted in the formation of the International Fire Engine Company, a firm that became part of American-LaFrance in 1904.

3-29
NYFD Engine Co. 30, circa 1909. This was the only triple company in the department. The company was disbanded in 1959. The building, designed by Edward Pierce Casey and opened in 1904, is now the New York City Fire Museum.

3-30
NYFD Engine Company No. 30, circa 1866. This was one of the first paid companies in New York City.

3-31
The hand-drawn hose cart or "jumper" of Mohawk Hose Company No. 1. Probably dating from the 1860s, this piece was photographed around 1907.

3-32
NYFD Engine Company No. 162 in 1907.

3-33
Everyone loves a parade. This shot, showing a NYFD steam engine, was taken around 1907. Note the straw boaters worn by the passengers on the rear step.

3-34
The NYFD in action at a fire in lower Manhattan around 1910. A hose tender is backed up to the steam engine, perhaps to deliver more coal to the engineer.

3-59
This gong is attached to the footboard of a horse-drawn steam engine from 1904. It was operated by pressing a pedal on the other side.

3-60
Instrumentation in the days of steam was elementary; the engineer watched the pressure gauge. The rookie polished the brass.

Amoskeag faced major competition from the Silsby Manufacturing Company of Seneca Falls, New York, which began making steam pumpers in 1856. The Silsby machines used rotary engines and geared rotary pumps based on a design by Birdsall Holley, rather than the piston or reciprocating pumps used by other makers. (Interestingly, Silsby also offered this pump sold separately as a feedwater pump for railroad locomotives.) According to its partisans, the rotary pump was simpler and supposedly produced less friction loss in the hose than piston pumps. The debate was never resolved, but Silsby was the only major manufacturer to offer the rotary pump. (The entire question became moot when gasoline-driven centrifugal pumps were developed.) Silsby produced more than a thousand steam fire engines before merging with several other firms in 1891 to form the American Fire Engine Company, later to become part of American-LaFrance.

M.R. Clapp, an employee of the Silsby company, left the firm and began building his own fire engines in partnership with a man named Jones. The firm of Clapp & Jones, headquartered in Hudson, New York, began producing steam engines in 1862. The company was quite successful, producing in all about 600 engines. Information about this important firm tends to be contradictory. It is possible that Clapp built at least one steam engine in 1860 while still based in Seneca Falls and before entering into his partnership with Jones. In any case, by the end of the 1860s the Clapp & Jones Company was producing excellent fire engines that utilized a characteristic rubber suspension system for the rear wheels. Almost all Clapp & Jones machines had vertical boilers, but the firm's designs were quite flexible. The pumps could be either double- or single-action, placed either horizontally or vertically. The air chamber was usually spherical (although some engines were made with an unusual cylindrical air chamber placed horizontally) and located at the front. The driver's seat was often, but not always, placed amidships. The apparent willingness of Mr. Clapp and Mr. Jones to accommodate the wishes of their clients may have had a lot to do with their success. The firm continued to produce outstanding fire engines for three decades, including a vertical engine and pump machine delivered to New York City's Engine Co. No. 53 in 1884. In 1891, the firm joined with several others to create the American Fire Engine Company. Immediately after the merger the new firm delivered an engine designed by Clapp & Jones to the New York City Fire Department. Christened Engine No. 44, this crane-neck machine was, in its beautiful design, decoration, and construction, the apotheosis of the steam engine.

After Alexander Latta retired in 1862, the firm of A.B. & E. Latta sold its original factory building in Cincinnati to the new firm of Lane & Bodley in 1863. The new owners continued on with the basic Latta concept of a self-propelled steam engine, utilizing the Latta-style square boiler with coiled tubes. (The tube system, known as a Buchanan coil, contained water within the tubes, as oppposed to other systems where the tubes were surrounded by water.) Lane & Bodley also carried on the three-wheel design pioneered by Latta, which al-

WYCKOFF
EXEMPTS
1904
STEAM TO GENL. SERV. PUMP
DOUBLE SPRING
MERICAN LA FRANCE FIRE ENGINE C
ELMIRA, N.Y.

3-59

3-60

3-35

3-36

3-37

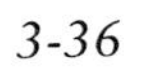

3-38

3-39

3-40

lowed the machine to turn around in its own length and thus gave it added maneuverability. However, the new firm also made substantial changes to the design to streamline and lighten it. Even so, the demand for this sort of machine never really developed, and eventually the firm gave up on self-propulsion. In 1868 Lane and Bodley sold their company to Chris Ahrens, who had been the firm's superintendent. Renamed the Ahrens Manufacturing Company, the firm moved into the large-scale production of steam engines. Giving up on the idea of self-propelled machines in favor of the more practical and popular Amoskeag-style engine, Ahrens continued to use a modified version of the boiler first developed by the Latta company. This excellent design allowed the engine to get up steam very quickly—in less than five minutes from cold water. This feature accounted for much of the success of the firm. Ahrens also added a small, hand-operated auxiliary pump next to the boiler, which allowed the engineer to pump manually until steam was raised and the circulating pump was in action. Most Ahrens engines used a vertically mounted single pump; the frames were either straight or crane-neck. All told, Ahrens Manufacturing Company produced more than 700 steam engines between 1868 and 1891, when the firm became part of the American Fire Engine Company.

In the 1860s a young man from Pennsylvania named Truckson LaFrance went to work at the Elmira Union Iron Works. (Young Truckson's original family surname was Hyenveux. When his French Huguenot family emigrated to America in the early 1700s, their surname was so unpronounceable to English speakers that they changed it, appropriately, to LaFrance.) In 1871 and 1872, Truckson LaFrance obtained several patents for improvements in rotary steam engines. The head of the firm was interested, and soon he and Truckson were making fire engines in a small way. By 1872, a significant new player entered the crowded, competitive fire engine business. The LaFrance Manufacturing Company of Elmira, New York, was named for its founder, Truckson LaFrance. Truckson's brother Asa joined the firm as traveling sales agent in 1876. By 1875 the first LaFrance steamer, utilizing rotary pumps for both the fire pump and the steam engine, had been delivered. The pumps used a new design featuring two bronze rotors contained in a bronze housing.

The fledgling firm had some shaky moments. In 1878 the company sent a steam fire engine to the Paris Exposition, in the hopes of stimulating worldwide sales. Unfortunately, they didn't know about French laws on boilerplate thickness, and the engine couldn't even be demonstrated. The expense of this venture nearly made the firm go under.

In 1880, the LaFrance Manufacturing Company was reorganized as the LaFrance Fire Engine Company. The change of name reflected a change of corporate policy. From the 1880 onward, LaFrance built fire apparatus strictly to the requirements of fire service. In effect, this meant that every piece of LaFrance apparatus was custom made to the specifications of the individual client. Although the LaFrance company was a late entry into the fire apparatus business, it quickly became a leader in the field. This rapid success was in large

3-35
A ladder company makes a practice run somewhere out west in 1909. The driver wears a bowler hat and seems to have lost the rest of his company.

3-36
A combination steam engine, using a piston engine and rotary pump, built by Silsby Manufacturing Company of Seneca Falls, New York, for the city of Chicago.

3-37
A steam engine from Silsby in the 1880s.

3-38
The steam engine of Citizen Engine Company No. 2, built sometime in the 1880s by Silsby Manufacturing Company.

3-39
A handsome parade hose carriage built by Silsby for the O.R. Packard Hose Company No. 1 is shown outside the plant in Seneca Falls, New York, in the 1880s. The price of $1,000 included the bouquet holders.

3-40
A Silbsy steam engine undergoes predelivery testing in the early 1880s. Four streams are aimed into the Erie Canal, conveniently close to the plant in Seneca Falls, New York. The exposure time for this photograph was 45 seconds. Despite the long exposure and four working streams, the exposure is quite sharp—an indication of the steadiness of the engine.

3-61
The New York Fire Insurance Patrol responds to an alarm in this print from the 1860s. Insurance patrols were sponsored by fire insurance companies. Their role was to save goods and property from the fire and to place salvage covers wherever possible to reduce water damage. Note that in this print the firemen go down a flight of stairs. Fire poles did not come into use until the 1870s.

3-64
The Pioneer, *a steam engine built by Reanie & Neafie in 1857. Note the unusually tall air chamber on the front-mounted pump.*

part due to the attention the firm paid to client needs.

In 1884, in response to demand, LaFrance built its first piston pumper. Eventually the firm offered piston pumpers in seven different size. The largest—extra first—had a capacity of 1,100 gpm. All the LaFrance piston pumpers had excellent independent sectional boilers with easy access to the valves. The design was very effective yet simple to operate and maintain. LaFrance continued to grow, producing about 500 steamers before it merged with the companies of the American Fire Engine Company and others to form the International Fire Engine Company in 1900.

A latecomer to the steam fire engine business was the Waterous Engine Works Company of St. Paul, Minnesota. This firm (which today continues to manufacture pumps for fire service) produced its first steam engines in 1888. Offered in five different sizes, the engines featured duplex piston pumps. The chief claim to fame of the Waterous fire engine operation, however, is the introduction of the first gasoline-powered engine—a momentous development that will be discussed in a later chapter.

Numerous smaller manufacturers also entered the steam engine field in the second half of the 19th century. Many produced only a handful of steam fire engines before sinking into obscurity. Others achieved a more significant level of production. For example, the Mansfield Machine Works of Mansfield, Ohio produced a number of attractive fire engines of different sizes in the years between 1883 and the late 1890s. Thomas Manning, Jr., and Co. of Cleveland began building engines in 1886, and produced about a hundred before being absorbed into the International Fire Engine Company in 1900. Manning engines bore a strong resemblance to those made by Amoskeag. They were fairly popular; the Detroit Fire Department had seven, and ten were sold to the Cleveland Fire Department.

THE CONSOLIDATION OF A FRAGMENTED INDUSTRY

The design of a steam fire engine was complex; building a steam fire engine was labor-intensive, slow, and expensive; the competition to sell engines was intense. By the end of the 1880s, it was becoming apparent to some successful fire-apparatus manufacturers that consolidation, not competition, would bring growth in the future. Led by the far-sighted Chris Ahrens of Ahrens Manufacturing Company, several firms merged together in 1891 to form the American Fire Engine Company, based in Cincinnati and Seneca Falls. In addition to Ahrens, the other firms involved were Button Fire Engine Works (founded 1834), Silsby Manufacturing Company (founded 1845), and Clapp & Jones (founded 1862). Although conceptually sound, the consolidation was fatally flawed. Each company continued to operate independently, and the expected savings and increased sales never really materialized—although managerial friction and disagreements did. Even so, the firm produced an excellent new

3-61

Bunk Room.
N.Y. Insurance Patrol.
Storage Room
At a Fire.
Saving Goods
188
HARPER'S WEEKLY.
March 4, 1876.
THE NEW YORK FIRE INSURANCE PATROL.—Drawn by I. Pranishnikof.—[See Page 191.]

3-64

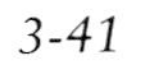

3-41

3-48

3-49

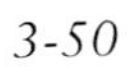

3-50

3-52

3-51

engine that quickly became a favorite among firefighters. Based on an Ahrens design called the American, the new design was named the Metropolitan. This design offered easy access to the valves and packing, making maintenance much simpler at the firehouse. (Metropolitan engines built by the American Fire Engine Company are easily identified today by the characteristic placement of the flywheel between the two vertical cylinders.) The firm also built several other models, including a combination hose cart and steamer called the Columbian. Fourteen of these steamers were built for the Chicago Fire Department in the 1890s; the model name comes from the Columbian Exposition, a world fair held in 1893 in Chicago. The fire companies equipped with these models can be considered the first single-piece units in the country. A crew of six was needed to operate a Columbian.

In 1901, the fire-apparatus industry was consolidated further by a massive merger. The companies making up the American Fire Engine Company, organized only a few years before, combined with two major engine manufacturers—the LaFrance Steam Engine Company and the Fire Engine Division (Amoskeag) of the Manchester Locomotive Works—and six smaller companies that made either fire engines or other firefighting equipment. The smaller firms were: Charles T. Holloway and Co. (founded 1879); Fire Extinguisher Manufacturing Co. (founded 1866); Rumsey and Co. Ltd. (founded in 1840); Gleason and Bailey Engine Works (founded 1845); Thomas Manning, Jr., and Co. (founded 1866); and Macomber Chemical Fire Extinguisher Co. (founded 1871). The new firm was called International Fire Engine Co.

The manufacturing plants of the assorted companies now making up International Fire Engine were scattered throughout the Northeast and Midwest. The new company consolidated production facilities at the LaFrance factory in Elmira, New York. In 1904, to further consolidate the merged firms, International Fire Engine Company changed its corporate name to American LaFrance Fire Engine Company, a name the firm would keep until 1927.

Around 1900 a new firm, the Nott Fire Engine Company of Minneapolis began producing steam fire engines featuring spiral-tube boilers, vertically mounted piston pumps and cylindrical, vertically mounted air chambers. By 1901 the company was renamed as the W.S. Nott Company. By 1904 it had developed a new, extremely efficient pumper that competed very successfully with the American LaFrance Metropolitan. The Nott Company achieved strong sales in a short time, but the end of the steam age was very near. Always innovative, Nott quickly turned to gasoline.

After the merger of his firm into American LaFrance, Chris Ahrens returned to the fire engine business in 1904. He joined with his sons Fred and John and his sons-in-law Charles Fox and George Krapp to found the new Ahrens Fire Engine Company, based in Cincinnati. Chris Ahrens designed a new model called the Continental. An excellent engine, it sold briskly on its own merits and on the reputation of its maker. In 1908 Charles Fox became president of

3-41
A major shipment of four Silsby steam fire engines and a hose wagon leaves the plant by rail in 1873.

3-48
A successor to the Uncle Joe Ross *was a 9,000-pound self-propelled steamer made by A.B. Latta. Purchased from money donated by residents, it was given the name* Citizens' Gift.

3-49
This three-wheeled Lane and Bodley steamer has a single pump and a cross-mounted engine. It dates from the 1860s.

3-50
The Lane and Bodley steamer used by Washington Fire Company No. 1 of Cincinnati, about 1860.

3-51
After building the Joe Ross, *Abel Shawk and Alexander Latta disagreed about the future of steam fire engine design and broke up their partnership in 1855. Shawk preferred a lighter model, such as the one shown here, although he continued to use the Latta coil-type boiler.*

3-52
The successor to the Lane and Bodley Company was Chris Ahrens, who in 1867 bought the Latta patents from his employers. Ahrens, one of the greatest innovators in fire engine history, made many improvements, starting with the use of four-wheel frames. This single-engine, straight-frame 1875 model is somewhat unusual because the flywheel is mounted transversely across the frame.

the company, and the name was changed to Ahrens-Fox.

As the nation entered the 20th century, the gasoline-powered internal combustion engine began to be a practical technology. At the same time, the steam-powered fire engine had reached a peak of efficiency that could not be refined much further. The firms making fire apparatus were quick to appreciate the possibilities of the gasoline engine. By 1914 American LaFrance had built its last steam engine; Ahrens-Fox built its last steamer in 1912.

3-53

3-53
Obliging firemen from Engine Company No. 43 in Cincinnati test an Ahrens-Fox engine before it is delivered. The main building of the Ahrens-Fox Fire Engine Company can be seen in the right rear background. A factory worker has been sent along to stoke the boiler.

3-54
Pheonix Ladder Company No. 1 of Cincinnati ran this immense, manually raised ladder.

3-63
Silsby steam fire engine #1013 is preserved today at the CIGNA Museum and Art Collection in Philadelphia. This engine dates from around 1889.

3-54

3-63

3-67
A jumper-type hose reel used by a Philadelphia company into the early part of the 20th century.

3-68
Reanie & Neafie built the Pioneer *for Philadelphia Hose Company No. 1 in 1857. It weighs 7,455 pounds.*

3-69
The Silsby Manufacturing Company built this steamer #1013 for Philadelphia around 1889. It is 11 feet long and nine feet high.

3-70
A "sidewalk" steamer, built by Sutphen in 1890 for Urbana, Ohio. Designed to be pulled along the sidewalk by its crew, this rig saved the town the expense of a horse. A jumper hose reel is next to the steamer. C.H. Sutphen, the founder of the firm (still operated today by members of the Sutphen family), is at left.

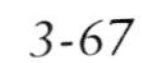

3-67

3-68

3-69

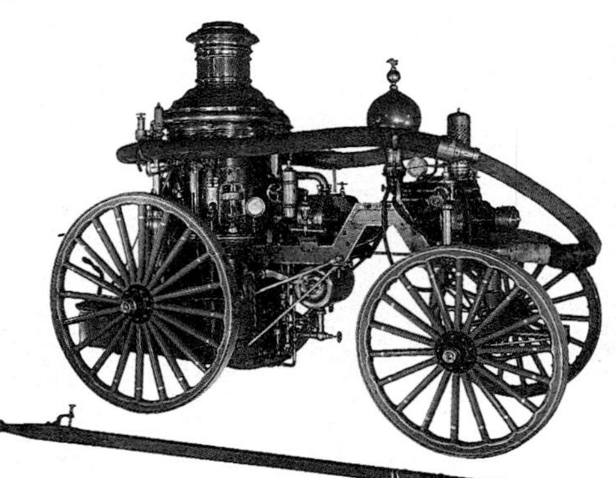

3-70

CHEMICAL WAGONS, AERIALS AND WATER TOWERS

The sight of a steam fire engine dashing out of the station house, drawn by a perfectly matched team of horses galloping flat out, was a thrill never forgotten by those who witnessed it. For decades, this picture exemplified the excitement and glamour of the fire service.

But to discuss firefighting only in terms of steam engines is to overlook some other important developments of the horse-drawn era.

CHEMICAL FIRE ENGINES

The combination of sulfuric acid (H_2SO_4) and sodium bicarbonate (known generically as baking soda and chemically as $NaHCO_3$) produces the gas carbon dioxide (CO_2). If sulfuric acid is added to water containing dissolved sodium bicarbonate and the resulting reaction is confined within a tank, the pressure of the carbon dioxide will build to the point that it can be used to force the contents of the tank out—a concept familiar to anyone who has ever shaken up a bottle containing a carbonated beverage. The idea of putting out a fire by controlling and directing the product of a sulfuric acid/sodium bicarbonate reaction was first developed in France in 1864 and used to make a portable fire extinguisher. The concept of a "soda-acid" fire extinguisher quickly caught on, and soon chemical fire extinguishers were common in Europe and America. By 1868, every fire house in New York City had at least one soda-acid extinguisher. When the alarm was for a minor fire nearby, a fireman with an extinguisher was dispatched on foot.

The fire extinguisher concept was carried a step further by the Boston Fire Department. A wagon carrying ten soda-acid fire extinguishers went into service in 1871 as Extinguisher Co. No. 1; two more companies, this time with 25 extinguishers apiece, went into service the next year.

Boston's bright idea never caught on elsewhere, however. This could well be because in 1872 the Babcock Manufacturing Company of Chicago developed

4-25
A head-on look at an 1890 hand-drawn American LaFrance chemical cart.

the first real chemical engine. Nothing more than a giant soda-acid extinguisher on wheels, this apparatus went into service with the Chicago Fire Department. It was the first of several thousand chemical engines the firm would produce, at first under the Babcock corporate name and later as the Fire Extinguisher Manufacturing Company (absorbed into the International Fire Engine Company in 1900 and thence into American-LaFrance). However, even after the merger, two basic types of chemical engines continued to be known by the Babcock trade name.

The Chicago Fire Department took delivery in 1871 of another chemical engine, this one designed and built by Charles T. Holloway. At the time Holloway was the first chief engineer of the Baltimore Fire Department. Holloway's life as a pioneer in the fire service was a fascinating and exemplary one. Chief of a junior volunteer company at 15, Holloway went on to organize the first hook and ladder company in Baltimore, and served as its president from 1850 to 1859. In 1859, when Baltimore went to a paid department, Holloway was the clear choice as first paid chief, a position he held until 1864. In 1868, however, Holloway returned to the Baltimore fire service as a fire inspector. In 1870 Holloway, originally a watchmaker, formed the Holloway Manufacturing Company in Baltimore. Its primary purpose was the manufacture of chemical fire extinguishers, using his own outstanding design. The firm was quite successful; as discussed in the previous chapter, it became part of the International Fire Engine Company in 1900 and later part of American-LaFrance.

The Babcock and Holloway designs were similar in concept and execution. The simplest Babcock design, known as the Champion, called for a horizontal tank with a handwheel at one end. The tank was filled with "soda water" (water mixed with sodium bicarbonate); a stoppered glass bottle containing sulfuric acid was suspended in a cage inside the tank. To discharge the engine, the fireman removed a safety pin and used the handwheel to rotate the tank, thus tipping the acid into the soda water and triggering the reaction. The discharge pipe was connected to a small-diameter hose—usually 3/4-inch, although 1-inch was used on the larger tanks. The manufacturer recommended tipping the tank occasionally as it discharged to get better working pressure and a better mixture of gas and water.

A more sophisticated design was the Champion-Babcock. Working on the same principle as the Champion, this model featured an agitator shaft with paddles. By rotating the paddle using an external handle (instead of tipping the tank), the chemicals were mixed more quickly and thoroughly. The result was higher pressure obtained sooner.

The Holloway model also operated along the same basic lines as the Babcock models. However, the Holloways offered an easily recognizable hood over the filling port for the sulfuric acid. Instead of rotating the tank to combine the acid and soda water, Holloway models used a lever near the filling hood to tip the acid jar over; an agitator shaft mixed the chemicals.

4-25

4-1

4-2

4-3

4-4

4-5

4-6

Babcock and Holloway competed for business, but Babcock took a clear lead to become the nation's largest chemical engine manufacturer. By 1886, more than two thousand Babcock chemical engines were in service.

Charles Holloway was successful in a smaller way. In part, this was because he remained on active duty as a fire engineer. In 1881, Holloway sold six double tank chemical engines to the newly formed Baltimore County Fire Department, for the tidy sum of $12,800. For their money, the county commissioners demanded Holloway's services as unpaid chief for two years. Over the years, additional Holloway provided additional chemical engines to the force, for a total of 29. Interestingly, from 1881 to 1894 (when the first steam pumper was purchased), the department consisted of nothing but chemical fire engines.

Competition from other manufacturers soon entered the chemical engine picture. F.S. Seagrave of Detroit joined the picture in 1891; Peter Pirsch & Sons Co. and Obenchain-Boyer of Logansport, Indiana also began manufacture of chemical engines. There were minor variations, particularly in the system used to put the sulfuric acid into action, but all worked along the same lines as the Babcock and Holloway systems.

A persistent myth about the fire-stopping ability of the chemical engine soon began. The extinguishing property of one gallon of chemical was variously claimed to be the equivalent of anywhere from ten to 40 gallons of plain water. How this obviously impossible belief arose, and why it lingered so long, are impossible to say. A confused notion that because carbon dioxide smothers flames it somehow helped the water have more effect may be behind the myth.

Myths notwithstanding, chemical engines were indeed extremely effective. They offered many advantages, particularly to smaller departments. Because the engine carried its own water, it was useful in areas without water mains. Most chemical engine rigs (including the larger, two-tank models) carried extra soda and acid and had a small, hand-operated water pump attached; the tanks could be recharged at the fireground if necessary. An average chemical engine developed pressure of 140 to 180 psi, although claims of 200 psi were sometimes made. Streams could easily be thrown 50 feet; using 3/4-inch hose, a 10-gallon tank could discharge at high pressure for at least 15 minutes. A survey conducted by fire insurance underwriters around 1896 revealed that chemical fire engines extinguished a large percentage of fires—as much as 90 percent in some cases. More typical was the response of Buffalo, New York, whose chief reported that chemical engines had put out 45 percent of all fires over a 19-year period. In Detroit, half of all fires over 19 years were doused by chemical engines; in Los Angeles over three years, 60 percent of all fires were doused with chemical engines. Oddly, Boston reported that only ten percent of its fires over 21 years had been extinguished using chemical engines—a figure that is surprisingly low. Chicago at one time had 13 two-wheeled and three four-wheeled chemical wagons in service; there were seven chemical engine companies in San Francisco in 1901.

4-1
A chemical wagon of the NYFD, sometime around 1900. The two vertically mounted chemical tanks can be seen by the rear wheels. The crew of seven shown here seems excessively large for a chemical company.

4-2
Kenmare, North Dakota Chemical Engine Company No. 1, ready for action in 1910. The hose is carried at the rear in a basket; the box above the horizontal tank carries tools.

4-3
Laurel Hose and Chemical Company No. 1, of York, Pennsylvania, in 1911. The two chemical tanks are carried horizontally slung under the wagon bed.

4-4
Members of the Wenatchee Fire Department Chemical Company No. 1 pose with their rig in 1911. The horizontal chemical tank is mounted transversely under the driver's seat.

4-5
The nicely appointed Union Hose and Chemical Company No. 3 of York, Pennsylvania is shown in this picture. The chemical tanks are mounted vertically between the driver's seat and the hose basket. The crews' helmets are neatly hung on the edge of the hose basket.

4-6
Double chemical tanks mounted transversely under the driver's seat are easily seen in this shot of Royal Hose and Chemical Company No. 6 of York Pennsylvania, taken in 1911.

4-24
This Holloway chemical engine from 1898 now resides at the Fire Museum of Maryland. It also carries a pompier, or scaling ladder.

4-26
In this view of the tank on an 1890 American LaFrance chemical cart, the brass-stoppered opening for the acid bottle is clearly visible at the top, just to the left of the hose basket. The discharge hose would be attached to the valve at the front of the tank, visible here between the spokes.

Chemical engines were available in numerous sizes, from two-man hand-drawn carts to horse-drawn double cylinder rigs and later to motorized vehicles. Particularly when compared to steam pumpers, chemical engines were lightweight and easy to maneuver—a typical hand-drawn, 45-gallon double-tank model might weigh only 750 pounds. Operating a chemical engine was extremely simple, and could be accomplished by just one fireman if necessary; maintenance was also simple. Partisans of chemical engines claimed that they caused less water damage than steam pumpers did. Finally, chemical engines were relatively inexpensive. However, there were also some serious drawbacks. Sulfuric acid (sometimes quaintly called oil of vitriol in old literature) and bicarbonate of soda were ongoing expenses, as were the acid bottles (especially those that had to be broken to release the contents). The general rule of thumb for the ratio of chemicals was one-fifth part sulfuric acid and two-fifths part sodium bicarbonate. Thus a 40-gallon tank would need 16 pounds of sodium bicarbonate. The chemicals were caustic and needed to be handled with care; any spills onto the painted or finished parts of the engine could mar the surface and needed to be removed quickly and carefully. A serious drawback was uneven mixing of the chemicals. When this occurred, the initial discharge from the hose contained a high level of acid, which corroded hoses, protective gear, and fabrics. Another significant drawback was that once the tank was activated, it had to be fully discharged, even if the fire was already out.

Two-wheeled, hand-drawn chemical carts usually had one or two tanks holding anywhere from 25 to 100 gallons. The hose was often carried coiled down in a basket, although hose reels were not unknown. The hose diameter was generally 1/2- or 3/4-inch (steam pumpers used 2 1/2-inch hose); 100 to 200 feet was usually supplied by the manufacturer.

Horse-drawn chemical engines usually carried two matched tanks, ranging in capacity anywhere from 25 to 100 gallons each. Horizontal or vertical placement of the tanks seems to have depended on the manufacturer in the earliest days. Later on, horizontal placement would become most common. Some 800 to 1,000 feet of hose was carried, almost always on a reel. By the 1890s, many hose carts and ladder trucks had been outfitted with one or two chemical tanks to make combination rigs. In these cases, the tank was often placed transversely under the driver's seat for easy access; the hose reel was usually mounted above and directly behind the driver.

As firefighting entered the motor age, chemical engines were still very much in evidence. Manufacturers such as American LaFrance (now incorporating both Fire Extinguisher Manufacturing Co. *and* Holloway), Obenchain-Boyer, and Pirsch produced motorized chemical cars. The rugged tanks could also easily be mounted by a mechanic onto any sort of motorized vehicle, such as a Model T Ford. Double and triple combination vehicles incorporating hose wagons or ladder trucks or volume pumps and chemical engines became common.

Although chemical tanks were very much a part of early motorized appara-

4-24

4-26

4-7

4-8

Champion Pneumatic Telescopic Electrically Insulated Aerial Hook and Ladder Truck, elevated against building.

4-9

4-10

4-11

4-12

tus, they were soon to become suddenly obsolete. Charles H. Fox, the mechanical genius behind Ahrens-Fox, invented the booster tank in 1913. This device used a small, gasoline-powered pump to discharge water from a small tank through a 1-inch hose. The booster tank was designed to fit where the chemical tank had previously gone. Boosters offered firefighters several major advantages over chemical engines. There was no cost for chemicals, and no corrosion damage. The pump operator discharged only as much water as was needed to extinguish the fire. In addition, the operator could control the pump pressure; once a chemical tank started to discharge, the operator had no control over the pressure at all. The obvious advantages of booster tanks quickly became apparent, and by the beginning of the 1930s chemical engines were gone from the firehouse.

AERIAL LADDERS

As cities grew larger and buildings grew taller in the period following the Civil War, firefighters began to need safe, efficient ladders that would reach the upper parts of multifloor buildings. The only available solution was clumsy, 75-foot extension ladders. These had to be manhandled into place by as many as nine firemen, and were dangerously likely to break.

A possible solution was offered by an extension ladder designed in 1873 in Milan, Italy by one Paolo Porta. The American rights to the ladder were sold to Mrs. Mary Bell Scott-Uda. The Scott-Uda aerial consisted of eight individual sections that were fitted together to raise the aerial to the required height (a maximum of 97 feet); one section was the base and was permanently attached to the frame of the wagon. To keep the ladder structure from overbalancing as it rose, a system of braces and counterweights was used; the ladder was raised using cranks and cog wheels.

Four Scott-Uda aerial ladders (as they came to be called) were ordered by the New York City Fire Department and delivered in the summer of 1875. The ladders were extensively tested after delivery, and a public demonstration of the new equipment was held on 14 September. Tragically, while seven men were on the ladder at different heights, the top section broke off, hurling three men to their deaths on the pavement 70 feet below. The new ladders never went into service, and in 1884 they were broken up and scrapped. A few more Scott-Udas were sold in other cities, but they too were seldom if ever used; the design and the Scott-Uda company disappeared. Aerial ladders in general came under a suspicious cloud in the minds of many fire chiefs.

A much more successful and pioneering manufacturer of aerial ladders was Daniel D. Hayes. Formerly a volunteer fireman in New York City, Hayes had sailed around the Horn in 1866 to deliver the first Amoskeag steamer to San Francisco. Finding the city to his liking, he stayed on and became chief mechanic for the fire department. In 1868, Hayes patented a design for an 85-foot wooden aerial ladder mounted on a turntable. The ladder was raised by using

4-7
Elaborate painting decorates Rescue Hose and Chemical Company No. 6 of York, Pennsylvania, in a photo taken around 1911. The cranks for the two chemical tanks can be seen protruding through the wagon bed above the front wheel.

4-8
Although chemical engines were effective for small fires, they were useless at big ones. In this photo from 1902 of a combined chemical engine and turret gun, the gun is in action.

4-9
A Champion pneumatic aerial ladder, elevated against a five-story building in this demonstration held around the turn of the century.

4-10
A tillered aerial ladder drawn by three beautifully matched horses gallops up the street in what was probably a demonstration in 1908.

4-11
In this advertisement from around 1860, a wooden aerial ladder is shown rescuing people from the roof of a four-story building. This apparatus, patented by Miekle & Carville of New York City, is raised by use of a windlass and rope; tormentors or supports are used to support the ladder and keep it from toppling backward. This is not the most practical of designs, and it is not known if any were ever produced.

4-12
Rex Hook and Ladder Company of the York, Pennsylvania Fire Department shows off its well-equipped wagon in 1911. Note the tillerman at rear, almost hidden by the overhanging ladder.

single horizontal worm gear; four to six burly firefighters were needed to turn the handle. The design, while sound, had two drawbacks: the horses needed to be unhitched before the men could reach the handle, and the tillerman rode in a compartment below the ladder, which restricted his view considerably.

The first Hayes aerial ladder was sold, not surprisingly, to the San Francisco Fire Department around 1872, at a cost of $3,000. A second ladder was sold to the department shortly after, and a few more were sold to other west coast cities.

Daniel Hayes never went into the manufacturing business himself, preferring to license his patent on a royalty basis. In 1877 he came to New York and arranged with the firm of Buckley & Merritt to manufacture and sell his aerial ladder in 80- and 100-foot models. An 80-foot model was demonstrated several times, but no significant sales ever resulted. In 1883 Hayes sold his patents to the LaFrance Fire Engine Company, who were considerably more successful in their marketing efforts. By 1886 the first LaFrance Hayes 85-foot aerial had been delivered to New York City (it became Ladder Co. No. 3). Within a few years, New York had eight Hayes aerials and Brooklyn had 14.

Other designs and manufacturers came along quickly. At Fire Extinguisher Manufacturing Company, a foreman named E. Steck received a patent for an aerial ladder in 1884. An aerial built on his design and called the "Babcock" was demonstrated in 1886. The Babcock was raised by means of a vertical worm screw on either side of the ladder's base; each worm screw was connected to a hand crank by bevel gearing. The tillerman on the Babcock sat above the ladder for better visibility. The first Babcock was sold to New Haven, Connecticut in 1887. Two years later, the firm had sold 24 aerials to various cities.

4-27
A double-cylinder Haywood chemical cart from 1913. The two openings visible here on the ends of the cylinders are for the handles of the agitator paddles, which are missing from this rig.

4-28
This double combination chemical and ladder was assembled on a Ford TT chassis in the Mt. Bethel, New York shop in 1921. The two horizontal chemical cylinders are in the rear; the tank directly behind the driver's seat is for gasoline.

A third major player in the aerial ladder market was the firm of Gleason & Bailey of Seneca Falls, New York. The company introduced its Dederick Aerial Ladder in 1895. The raising mechanism for the Dederick consisted of raising arms connected to bronze cables, which were in turn connected through pulleys to a drum at the base of the ladder. The drum was revolved using crank handles attached to a chain-link reduction drive. Raising a Dederick aerial was thus quick and easy.

The mechanical difficulty of cranking an aerial up by hand led to the search for new methods. In the late 1880s, Chief E.F. Dahill of New Bedford, Massachusetts invented a system using compressed air in pistons. The Dahill Air Hoist, as it was called, was soon offered by most aerial manufacturers. In 1902 Frederick S. Seagrave, a ladder manufacturer in Detroit, offered a patented spring-operated hoist system to raise the aerial. Twin lifting springs raised the main part of the ladder; cranks and a worm screw extended it. This system, which came to be known as "manual spring-assist," had a major impact on the aerial business. Soon most makers were offering manual spring-assist. American-LaFrance introduced a spring-assist aerial in 1904. The mechanism was complex, and featured a hydraulic cylinder to control the extension of the lad-

F.D.
ST.JAMES
FIRE DEPT
NO TRACTOR
TRAILERS
IN THIS
BAY
MT. BETHEL
VOLUNTEER
FIRE CO NO 1

4-27

4-28

4-13

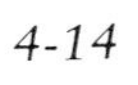

4-14

4-15

4-16

4-17

4-18

der—the first step toward the all-hydraulic systems that would later predominate. Other manufacturers offering spring-assist raising included Pirsch and Ahrens-Fox. Later improvements led to combinations of spring and compressed air, eliminating the hand-cranked worm screw assembly. By the 1930s, when gasoline engines had taken over in the firehouse, aerials were lifted hydraulically using a power take-off from the transmission.

WATER TOWERS

Aerial ladders were meant chiefly for ventilation, rescue operations, and the like. To run hose lines several stories up along ladders of any sort was risky, and ground-based deluge devices reached no higher than three stories. The need for apparatus that could pour a powerful stream of water into the upper stories of a building—a water tower—was clear. The first working attempt to deal with the problem was a "hose elevator" patented by one Mr. Skinner of Chicago in 1869. Resembling a modern snorkel more than anything else, this apparatus elevated the hose in a sort of bucket; it could reach a height of 40 feet. Skinner's hose elevator went into service with the Chicago Fire Department, and was used in the Great Chicago Fire of 1871. The Chicago Fire Department liked the idea of hose elevators so much that it purchased another, this time a model designed by a Mr. Knockes and modified for fire service. This model was also used at the Great Chicago Fire; both hose elevators were scrapped by 1873.

The first true water tower was invented in 1876 by Albert and Abner Greenleaf and John B. Logan, all of Baltimore. The name aptly given by the partners to their invention was "Portable Stand-Pipe or Water Tower." The Greenleaf water tower consisted of three sections. A base pipe was permanently mounted on the floor of the wagon; a straight pipe or mast was attached to the base; finally, a 1 1/2-inch diameter extension mast pipe with a built-in nozzle tip was attached to the straight pipe. The pipes were raised by a hand-cranking an assembly of gears on a trunnion. A hand-cranked wheel at the rear operated the leveling system. A cable braced the main mast against back-pressure.

A 50-foot Greenleaf water tower was successfully demonstrated in June of 1879, and was loaned to the New York Fire Department in July of that year. The inventors received their patent in 1880. In 1881, New York purchased the loaned water tower outright for $4,000. The Greenleaf firm delivered a second 50-foot water tower to New York in 1882. This rig had a larger extension mast pipe (2 1/2 inches) and two interchangeable nozzle tips. It also had an early version of a deck pipe mounted on the wagon. This water tower had the misfortune to overturn while undergoing acceptance tests. With modifications to the counterweight system of bracing, the tower was nonetheless accepted for service. One more Greenleaf, this time a 50-foot water tower for Boston, was delivered in 1882. In 1883, Greenleaf and Logan sold their patent rights to the Fire Extinguisher Manufacturing Company (FEMCO) of Chicago. However,

4-13
The Seagrave American Automatic aerial truck of Engine Company 7 poses in front of the fire house on Duane Street in New York City. The horses that originally pulled this spring-raised unit were replaced by a Christie front-drive tractor around 1917.

4-14
Cincinnati Ladder Company No. 2 in 1917. Note the hard rubber tires on the Ahrens-Fox tractor. The aerial ladder is a 1917 Seagrave 85-footer.

4-15
This NYFD 65-foot FEMCO water tower went into service in 1898, originally drawn by three horses. It was fitted with the Couple Gear tractor shown here in 1911. The Couple Gear Freight Wheel Company of Grand Rapids, Michigan, built combination gasoline/electric vehicles. The gas engine powered individual electric motors attached to each wheel. This arrangement was a primitive effort to solve the transmission and differential problems of early motorized apparatus. With various tractors attached, this water tower remained in active service until 1957.

4-16
NYFD Water Tower No. 2, part of Battalion 2, parked outside the firehouse at White and Lafayette streets. This clear shot shows many details of the 65-foot FEMCO water tower and the tractor. Note the right-hand drive.

4-17
NYFD Water Tower No. 3, a 65-foot American LaFrance model from 1907, was motorized with a Garford tractor. As shown in this photo, the tractor was crushed by falling debris at a fire in June of 1918. Water towers could throw streams so powerful that they punched holes in brick walls.

4-18
NYFD firemen raising an aerial ladder sometime around 1910. Note the double bucket seats and right-hand drive of the Seagrave Model AC "buckboard" tractor. This unusually sharp photo shows the details of the hand-operated raising mechanism and hydraulic cylinders.

4-29
A good look at the hose reel and siren on a chemical wagon—in this case, on Mt. Bethel, New York's 1921 Ford TT chassis.

4-30
The Mt. Bethel, New York, Fire Department purchased this chemical wagon in 1890. A group of beautifully polished hand-held brass fire extinguishers is in the wagon. The hump in the center of the tank held the acid bottle. The handle for the agitator blades can be clearly seen at left.

FEMCO built only one water tower along the Greenleaf lines, a 50-footer delivered to New York City in 1884. Water towers were an expensive investment for a city, and it is to be expected that they sell slowly. However, the real reason FEMCO failed to exploit its newly acquired patent is that it very quickly became obsolete.

In 1886, George C. Hale, chief of the Kansas City, Missouri Fire Department and, not coincidentally, president of the Kansas City Fire Department Supply Company, invented a manually raised water tower. He received a patent on the design and built a 50-foot prototype that year; a second, 75-foot model was built in 1888. Both went into service with the Kansas City firemen. Hale's real breakthrough came in 1889, when he developed a combination chemical-hydraulic lifting system. Water pressure was used to force the soda-acid mixture from the tank into two lifting cylinders. The chemical reaction took place in the cylinders, and the resulting carbon dioxide gas moved pistons that engaged a gear assembly. Hale built 20 water towers using the chemical raise system. (The chemical assembly was provided by FEMCO.) From 1886 to 1892, Hale had a virtual lock on the water tower business; in all, he built 38. But by the time Hale delivered a 55-foot water tower to Syracuse, New York, he was facing aggressive competition, particularly from a new design manufactured by FEMCO.

Having realized that the Greenleaf design was inadequate in the face of Hale's inventions, FEMCO fought back by developing an entirely new design. In 1893 FEMCO introduced the Champion model, based on a patent held the firm's general manager, Ernst Steck. (The model name capitalized on the firm's famed chemical engines of the same name.) The Champion was superior to the Hale design in several ways. It featured an excellent nozzle (salvaged from the old Greenleaf design). More importantly, the mast was placed on a turntable instead of trunnions; this meant that the tower could operate at an angle to vertical. The Hale water tower had the serious competitive disadvantage of operating only in the vertical.

Champion water towers were raised by hand. Two men were needed on each side of the mast to turn the handles placed there. The handles connected to bevel gears. The nozzle pipe was extended using a hand crank connected to gears that moved a cable. The first six Champion towers had the turntable placed at the front of the wagon; the nozzle hung over the rear. The 15 Champions built after 1894 had a rear turntable; the nozzle end projected over the horses in front. All Champion water towers were designed with a very low center of gravity. Theoretically, this provided more stability, but Champion towers were prone toward overturning—indeed, nearly every Champion ever built overturned eventually.

The first sale of a Champion water tower was a 65-footer to Detroit in 1893. The unit saw front-line action for 30 years. Several Champion towers, including the second ever built, were exported to Canada. Montreal purchased a 65-foot front-turntable tower in 1893, and added a 75-foot rear-turntable

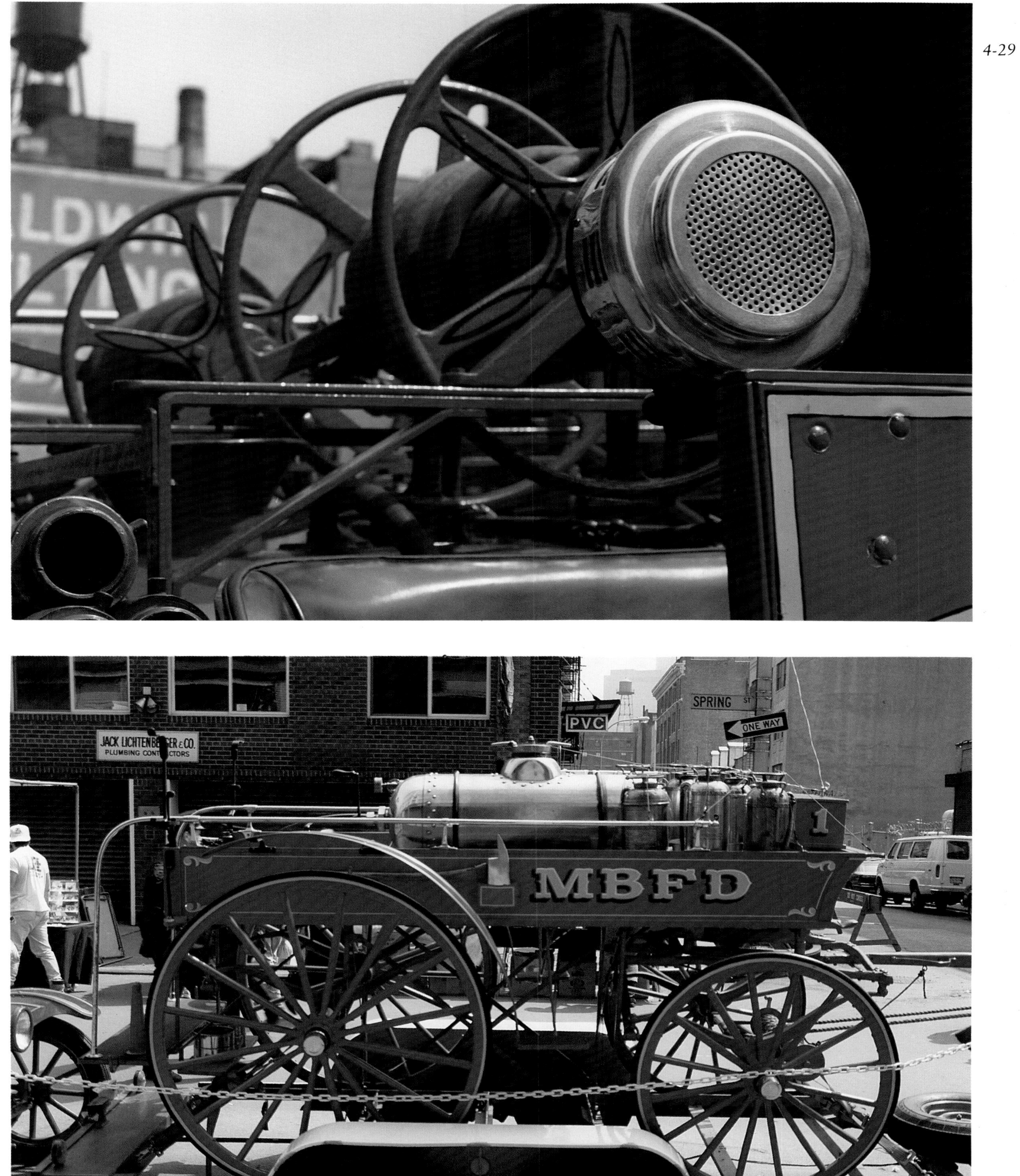

4-29

4-30

4-19

4-20

4-21

4-22

4-23

model in 1899. Toronto received a 65-foot tower in 1897; this tower saw considerable service, and actually survived intact and unmotorized for nearly 50 years before being scrapped sometime in the 1940s.

Sales of the Champion water tower were brisk right from the start. In 1894 George Hale, tired of the competition, sold his patents to FEMCO and retired from the apparatus business. Now in possession of the Greenleaf, Hale, and Steck patents, FEMCO started offering a new water tower design. The first of these was delivered to New York City in 1895. The new model incorporated the Champion-style nozzle and the Hale chemical-hydraulic lifting system.

For a brief while, FEMCO enjoyed a near monopoly on water tower sales. In 1898, a flicker of competition arose when a new water tower design arrived on the scene from San Francisco. A fire department mechanic named Henry H. Gorter came up with the idea of using a water pump motor as the raising mechanism, eliminating slow and exhausting hand-cranking. (Unfortunately for firemen, this concept never caught on with other manufacturers.) Gorter's design also featured a new and improved nozzle that connected to the supply pipe via a ball-and-socket joint. This allowed the nozzle to rotate 360 degrees in the horizontal and almost 180 degrees in the vertical. The increased flexibility this offered firefighters was immense. Gorter water towers were also very stable, with a low center of gravity and a bracing system that kept the towers from overturning.

Gorter's first tower was 65 feet long; it was delivered to San Francisco in 1898. A second tower, this one 76 feet long, was delivered to San Francisco in 1901; it was the tallest water tower ever built. In 1905 Gorter sold a water tower to Los Angeles.

Excellent as the Gorter design was, the firm was never any serious competition to FEMCO, and Gorter seems to have quietly faded from the scene after 1905. In 1926, however, the San Francisco Fire Department decided it needed two new aerials. Dusting off the original Gorter plans from 1898 and 1901, the department contracted with the Union Machine Company to produce two 35-foot water towers. The only concession to changing times was mounting the towers on Mack AC Bulldog motorized chassis.

In 1900 FEMCO and several other apparatus manufacturers merged to form the International Fire Engine Company, which became the American-LaFrance Fire Engine Company in 1904. This meant that the Greenleaf, Hale, and Steck patents were now held by American-LaFrance. The first water tower to be produced under the new management was a 55-footer delivered to Atlanta in 1904. That same year, the Seagrave Company (now the only real competition) introduced a new spring-raise mechanism. The first two Seagrave water towers to use the new assembly were built in Canada by William E. Seagrave, a relative of founder Frederick Seagrave. The first, 65 feet long, was delivered to Winnipeg in 1905; the second, also 65 feet long, went to Montreal in 1906. Seagrave's first American-built water tower went to New York City in 1907.

4-19
A wall collapse has damaged the NYFD apparatus in this undated photograph, probably from around 1912. The tractor-drawn hose tender looks to be total loss, while the FEMCO water tower seems damaged but more or less intact.

4-20
An unidentified NYFD FEMCO water tower in action in 1917.

4-21
The deck nozzle of a Seagrave 1907 65-foot duplex water tower is demonstrated in New York City. This tower was drawn by a three-horse rig.

4-22
Water Tower No. 3, Engine Company 31, of the New York Fire Department. This 1900 FEMCO was 65 feet long. The details of the Couple Gear tractor attached to the rig in 1912 can easily be seen in this clear shot.

4-23
FDNY Water Tower No. 1, Engine Company 31, a 65-foot FEMCO from 1898, in front of the fire house on Lafayette Street. The tormentors or jacks used to help keep the tower from toppling over due to back pressure can be seen here.

4-31

4-32

4-33

Responding to Seagrave's challenge, in 1909 American-LaFrance produced the last Champion and introduced a new line of water towers known as American Automatic. These rigs were offered in a variety of configurations meant to accommodate all budgets and firefighting needs. The major selling point of the American Automatic was the spring-raised lift. The first of the new models was sold to Knoxville, Tennessee in 1909. And as the horse-drawn age came to close, an American Automatic was sold to Houston in 1912. It was the last horse-drawn water tower ever built.

4-34

4-31
In this close-up view of the tank on a chemical cart, the discharge valve can be easily seen at the top.

4-32
The business end of a 1924 American LaFrance chemical truck. The interconnecting valves let the firemen discharge the tanks singly or together.

4-33
The grille end of a 1924 American LaFrance chemical truck. The chassis is a well-preserved Brockway Torpedo.

4-34
An American LaFrance chemical wagon with double Champion tanks, belonging to the Valdusta, Georgia, Fire Department around 1912. The double arrangement allowed one tank to be discharged while the other was being filled with water.

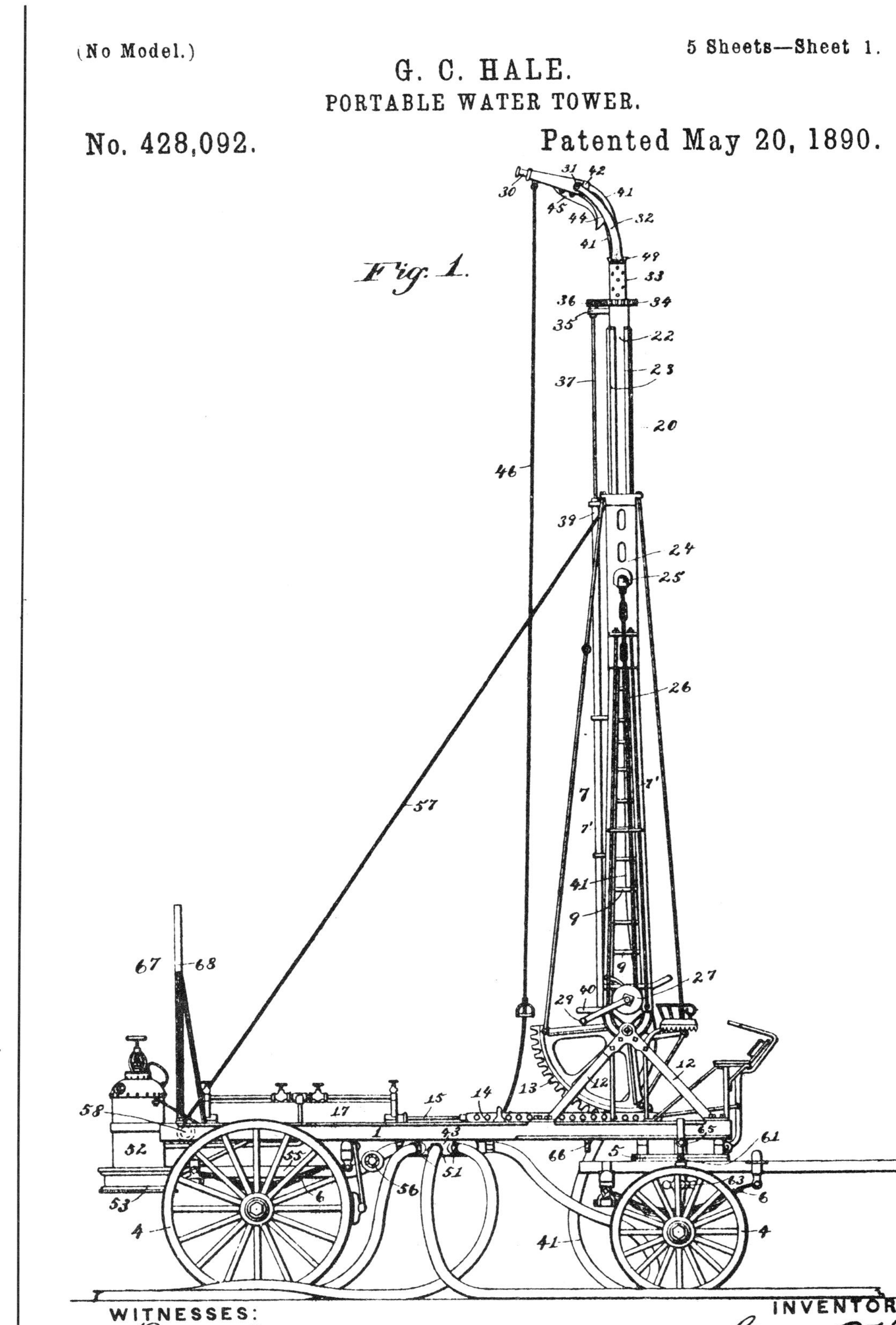

4-35

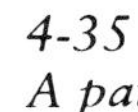

4-35
A patent drawing from 1890 for George C. Hale's hydraulically raised 45-foot water tower. A soda-acid tank, similar to those used in chemical carts, is mounted on a platform at left. When the tank was activated, the hydraulic pressure raised the tower.

4-36
Henry H. Gorter of San Francisco challenged the FEMCO lock on the water tower market with this model, patented in 1900. The tower was raised using a water motor.

4-37
The Gorter water towers featured a mast nozzle using a ball-and-socket joint. This concept allowed the nozzle to be positioned at any angle in the vertical, an innovation that gave much more flexibility and actually long outlived the water towers themselves.

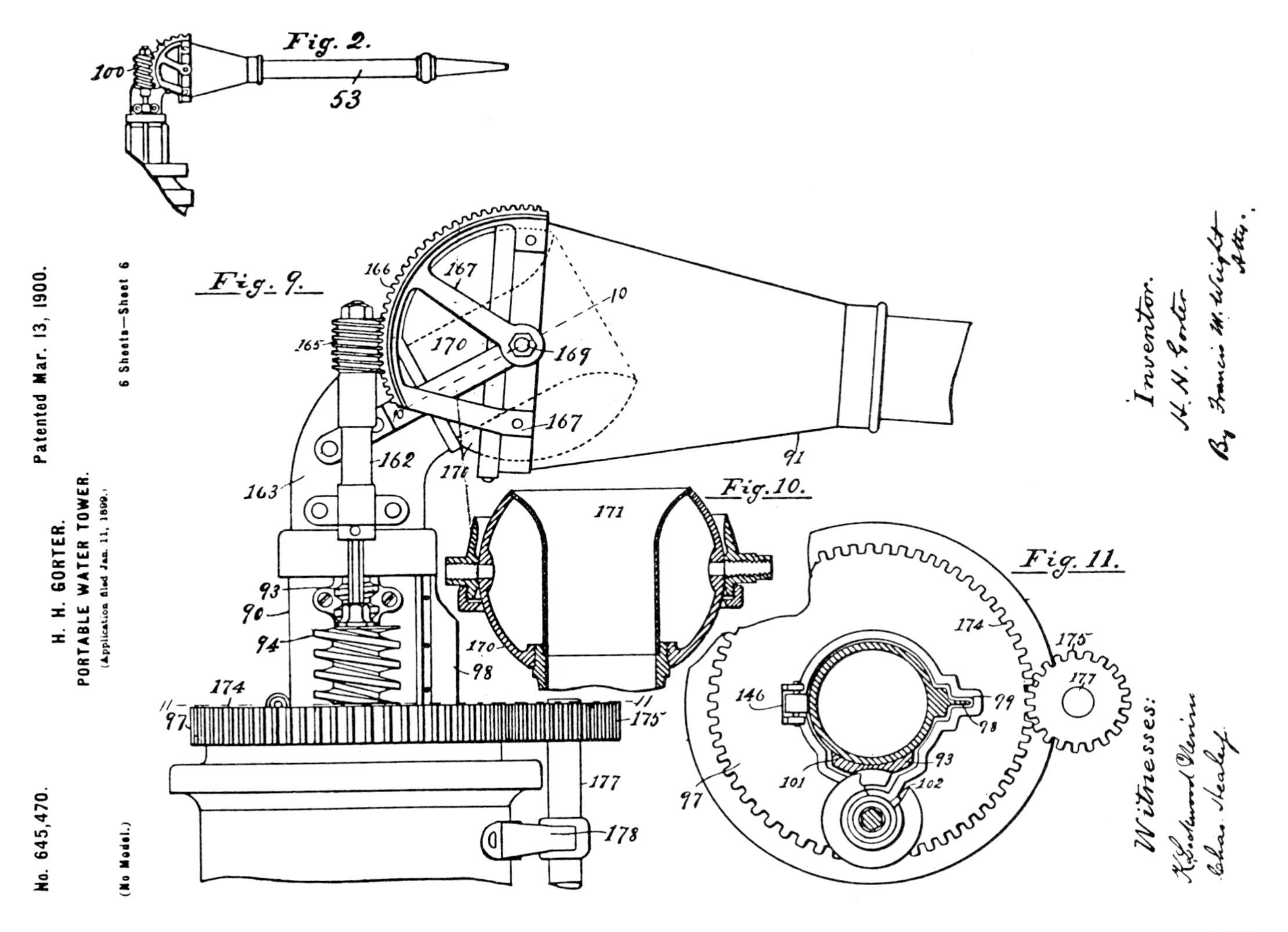
Patented Mar. 13, 1900.
H. H. GORTER.
PORTABLE WATER TOWER.
(Application filed Jan. 11, 1899.)
(No Model.)
6 Sheets—Sheet 6
No. 645,470.
Fig. 2.
Fig. 9.
Fig. 10.
Fig. 11.
Inventor.
H. H. Gorter
By Francis M. Wright Atty.
Witnesses:

4-37

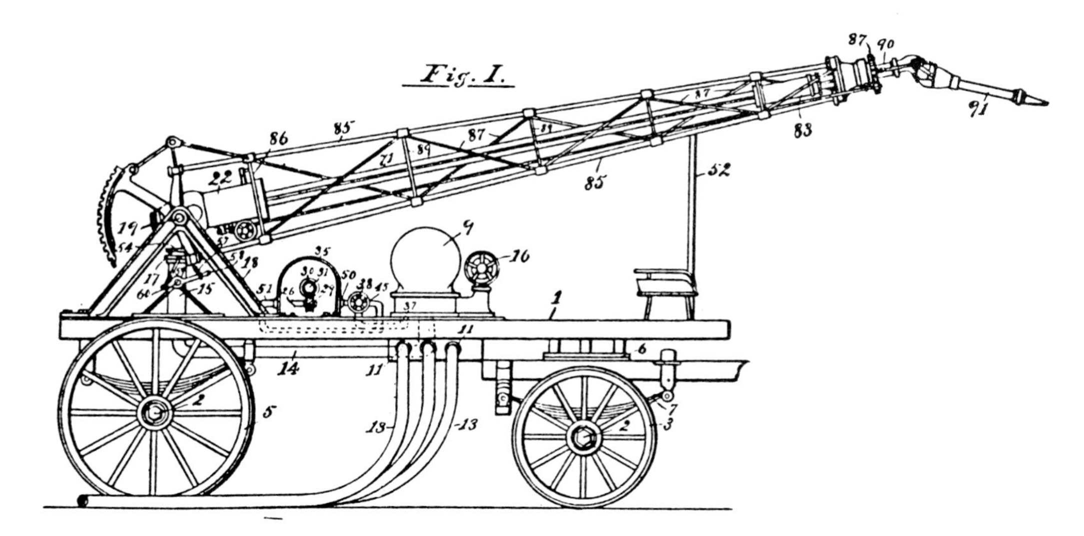
Fig. I.
Patented Mar. 13, 1900.
H. H. GORTER.
PORTABLE WATER TOWER.
No. 645,470.

4-36

4-38

4-39

4-38
This early version of the Mack Model AB dates to 1915. This chemical unit served Bridgeport, Connecticut.

4-39
A Mack Model AB pulls this aerial ladder in a shot from around 1916.

HORSEPOWER TO GASOLINE

By the beginning of the 20th century, a typical American city could congratulate itself on having an up-to-date, professional fire department boasting steam-powered pumpers, hook and ladder trucks, hose wagons, chemical engines, miscellaneous auxiliary rigs such as coal tenders, and perhaps a water tower—all drawn by horses. Change was in the air, however, and soon much of that modern equipment would be obsolete.

THE INTERNAL COMBUSTION ENGINE

In 1824 a French scientist named Sadi Carnot propounded the basic principles of the internal-combustion engine—that is, an engine that burns a mixture of fuel and air in such a way that the resulting hot gases exert direct force on pistons, enabling the engine to do useful work. Carnot's work had a powerful influence on another Frenchman, an engineer named Alphonse Beau de Rochas, who developed the concept of the four-stroke engine in 1862. Modern gasoline-powered vehicles operate using Rochas' idea, but today the four-stroke cycle is better known as the Otto cycle, after a design developed in 1876 by the German inventor Nikolaus August Otto. (Even though Rochas later challenged Otto's patents in court and won.) Basically, the Otto cycle consists of a four-stage sequence of events that occur within a cylinder fitted with a movable piston. In the first stage, intake, the piston descends within the cylinder and the resulting suction draws an explosive mixture of gasoline and air through an intake valve into the cylinder. In the second stage, compression, the fuel mixture is compressed as the piston rises again. As the piston nears the top of its path in the third part of the cycle, power, the compressed fuel mixture is ignited by an electric spark from the spark plug. The burning fuel forces the piston down. In the last stage, the piston rises again and forces the exhaust from the burned gasoline of the power cycle out of the cylinder through an exhaust valve. The four piston strokes result in two turns of the crankshaft. Otto patented his design in Germany and received a US patent in 1877; production of Otto engines in America began in 1878.

5-1
Two generations of NYFD apparatus are visible in this photo from around 1915. A steam engine drawn by a Christie front-drive tractor and a Seagrave "buckboard" are to the left; a more modern motorized Seagrave with turret gun is at the right.

5-2
The Hopewell Fire Department shows off its combination chemical wagon in this photo, taken around 1910.

As with steam engines, the first gasoline engines were meant as stationary engines for operating machinery such as looms. However, in 1885 Carl Benz, a German engineer, developed the first true gasoline-powered, internal-combustion automobile. This three-wheeled vehicle was powered by a two-cycle, one-cylinder engine. The first public appearance of the new vehicle took place on 3 July 1885 in Mannheim. Benz reached a speed of 15 kilometers per hour. At a later test drive that autumn, carried away perhaps by such heady speeds, Benz forgot to steer and crashed the vehicle into a brick wall by his house. Never daunted, by 1888 Benz employed a work force of 50 to produce his tricycle cars.

Working independently of Benz was another German named Gottlieb Daimler. After working with Nikolaus Otto for more than a decade to develop the four-stroke gasoline engine, Daimler left the firm, primarily because Otto saw the gasoline engine as suitable only for stationary use. Daimler and another former Otto employee, Wilhelm Maybach, began manufacturing automobiles using the first high-speed internal-combustion engine (900 rpm, as compared to 250 on the Benz engine). They successfully mounted the engine on a bicycle on 10 November 1885, and powered a four-wheeled modified carriage with it in 1886.

The credit for creating the first true purpose-built automobile (not a modified carriage) goes to a French engineer named Emile Levassor, a Daimler licensee. In 1889 Levassor developed the basic concepts that still govern modern automotive design: front-mounted engine, steerable front wheels, clutch, transmission, drive shaft, differential, and rear-wheel drive.

By the 1890s, automobiles, as the newfangled, self-propelled vehicles came to be called, were seen on American roads. Not all of these vehicles were powered by gasoline. Steam-powered automobiles (such as the famous Stanley Steamer) and electric vehicles (powered by batteries) were also seen. In 1898 there were more than 50 different American automobile manufacturers. The first commercially successful American automobile came along in 1901: the original three-horsepower, curved-dash Oldsmobile, designed and built by Ransom Eli Olds. In 1901 Olds sold 425 vehicles; in 1904 he sold 5,000. Olds' success inspired numerous others to enter the field, and by 1904 there were nearly 250 American automobile manufacturers, including a fellow named Henry Ford. The internal-combustion engine had arrived, and massive change was once again on the way for the fire service.

MOTORIZED FIRE ENGINES

The concept of a self-propelled fire engine was not a new one to the American fire service. As discussed earlier, steam-powered self-propelled vehicles of varying levels of practicality had been put into service over several decades. Indeed, the important concept of the differential gear was first developed and applied to a self-propelled steam engine in 1872. All these machines were inherently

5-1

5-2

slow and cumbersome, however, and horse-drawn apparatus was almost always more efficient.

On the other hand, memories of the stubborn but pointless resistance to the introduction of steam pumpers in the 1860s and 1870s were with every fire chief. While the fire service remained fundamentally extremely conservative, many firefighters were cautiously receptive to the potential of the horseless carriage.

The first automobile to be put into fire service was not powered by gasoline. It was an electric runabout purchased in 1901 for the use of the chief engineer of the San Francisco Fire Department. The second automobile in fire service was steam-powered: a Locomobile purchased personally in 1901 by Chief Edward F. Croker of the New York Fire Department. By 1904, the department was convinced of the automobile's practicality. Chief Croker was presented with a gasoline-powered American-Mercedes touring car. This vehicle was powered by a four-cylinder engine and developed an amazing 24 horsepower; it also featured tufted leather upholstery.

THE FIRST GASOLINE PUMPERS

In 1844, Mr. C.H. Waterous founded the Waterous Engine Works Co., Ltd., in Brantford, Ontario. The original intention was to manufacture engines of whatever sort required. This eventually came to include steam fire engines. By 1881, Waterous fire engines were in demand, and a branch was opened in Winnipeg by Fred L. and Frank J. Waterous, the twin sons of the founder. Soon Fred and Frank were distributing their own fire apparatus and also that of other manufacturers. The company had grown beyond the capacity of the Winnipeg facility. Across the border was the rapidly growing town of St. Paul, Minnesota, conveniently situated near rail and water transport and eager to attract manufacturing enterprises. In 1886 the Waterous brothers moved their operations, building a new factory on land given to them by the city. They began making steam pumpers the same year. By 1890 their small (250 gpm), lightweight (2,200 pounds) steamers, designed expressly for the rural market, were selling well. In 1898, the Waterous firm came up with an amazing innovation: the first gasoline-powered pumping engine. Mounted on a chassis very similar to that used for steam engines, the engine was designed to be pulled either by hand or by horse. The actual operation of the pump was a bit tricky, if not downright dangerous. Engine ignition was achieved through the "hot tube" method. A platinum tube was screwed into the combustion chamber of the engine. The tube was heated with a blowtorch. Compression forced the vaporized gasoline (or kerosene—the engine ran on either) through the tube into the chamber, where it exploded to start the engine. (This alarming method was also used to start early automobiles, and was later replaced by the hand crank and—after 1912—by the electric self-starter. It was the development of the easy-to-operate self-starter that allowed many women drivers to take to the

5-37
This well-preserved 1923 Oshkosh combination features four-wheel drive.

5-49
The first motorized apparatus of the Uniondale, New York Fire Department was this 1924 Larrabee triple combination.

OSHKOSH

5-37

85
BROOKSIDE
UNIONDALE FIRE DEPT.
1924
LARRABEE-CHILDS
OUR FIRST MOTORIZED
FIRE APPARATUS
UNIONDALE
FIRE DEPT.

5-49

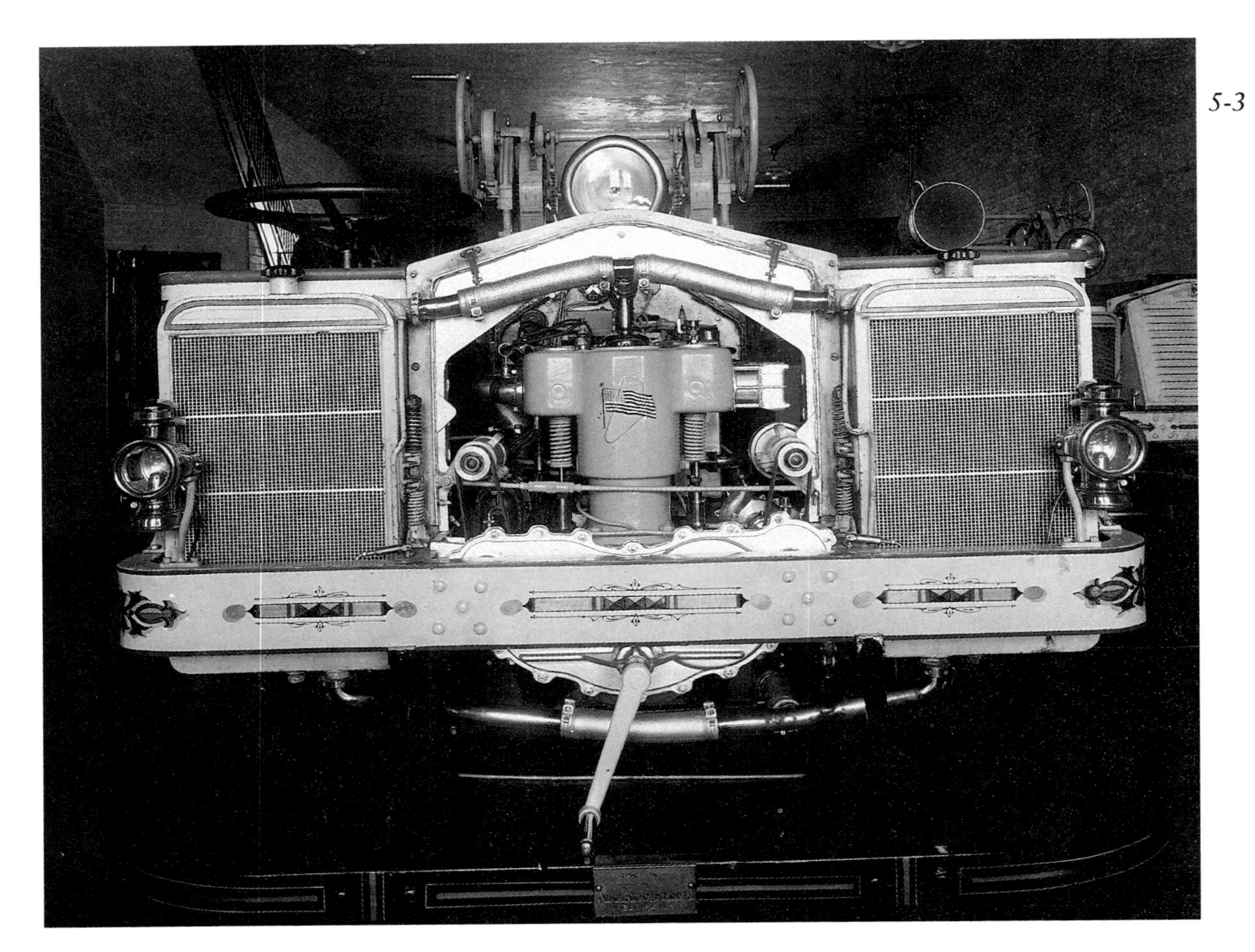

5-3

5-4

5-5

5-6

road.) A flywheel moved the piston, which was connected by a clutch to a 300-gpm rotary pump, through the exhaust, intake, and compression strokes. Primitive and potentially dangerous, this mechanism was also inexpensive. It sold well through 1907 to rural communities that wanted the efficiency of a steam pumper but could not afford the expense.

The credit for the first ever motorized fire apparatus should perhaps go to American LaFrance, which delivered a steam-propelled combination hose and chemical wagon to New London, Connecticut in 1904 (although this unusual rig should be considered more of an aberration than a development). However, in 1906 Waterous achieved another significant breakthrough when it mounted a gasoline-powered 300 gpm pump on the back of an automobile chassis. This dual-engine rig was the first self-propelled, gasoline-powered pumping engine in North America. It was also the first fire engine to have left-hand drive—a feature that was a still long way from becoming standardized. It was delivered to the Radnor Fire Company of Wayne, Pennsylvania. Sometime around 1907 Waterous sold a single-engine pumper to the Alameda, California fire department. The four-cylinder engine propelled the rig and also powered the pump, producing 600 gpm.

The New York Fire Department took a significant step forward in 1909 when it purchased a Knox high-pressure hose car—its first piece of motorized apparatus. As Fire Commissioner Waldo Rhinelander noted in his annual report for 1910, the rig cost only $85 a year in gasoline (which cost about 12¢ a gallon in those palmy days), oil, and repairs. By contrast, a team of three horses cost the city $660 a year. In 1910, the Detroit Fire Department put its first motorized rig into service.

By 1911, the first completely motorized fire department in America was Savannah, Georgia. The battery consisted of seven American LaFrance pumpers, four combination chemical and hose carts, and one straight chemical engine. And by 1922, around the country, with much ceremony and many a tear, the last fire horses were retired: every fire department was fully motorized.

5-3
An up-close view of the front end of an American LaFrance Type 16 straight-frame aerial ladder truck. Twenty-five of these combination gasoline/electric rigs served the NYFD starting in 1913. The four-cylinder engine generated electricity, which was used to drive the front wheels.

5-4
Assistant Ranger Farley used this Maxwell runabout to get to fires in Sierra National Forest, California, in 1912.

5-5
New York Fire Department equipment at a three-alarm fire in 1913.

5-6
NYFD Engine Company No. 93 poses on Washington Bridge (in the Bronx, not the George Washington Bridge of today) around 1913. The steam engine is drawn by a Christie tractor.

THE EARLY MANUFACTURERS

In the early years of the automotive industry, numerous small, untried companies entered a wide-open new market. For example, in 1906 the first automobile chemical wagon was produced by the Knox Automobile Company of Springfield, Massachusetts. This vehicle offered two 35-gallon chemical tanks and achieved a top speed of 40 miles per hour. Knox would remain a producer of fire apparatus for several decades.

Of the companies that entered the business, some, such as Mack and Pirsch, remain in business today. Others, such as Stutz and Obenchain-Boyer, were moderately successful but eventually succumbed to market forces. Still others, including companies with such unlikely names as the Tea Tray Manufacturing Company of Newark, New Jersey and the Davis Sewing Machine Company of

5-50

5-50
Under the hood of this 1924 Larrabee is a six-cylinder Continental engine.

5-51
This nicely preserved Seagrave Suburbanite from 1929 is owned by the North Adams, Massachusetts Fire Department. The tires are modern replacements.

5-52
A 1924 American LaFrance rig owned by the Columbia, South Carolina, Fire Department. Note the box-like extension on the front of the rear fender; this accommodates the chain drive.

Dayton, Ohio, produced a handful of interesting fire engines and eventually faded into oblivion. The Tea Tray Company, for example, is credited with making the first triple combination truck (hose, pump, and chemical tank), which went into service with the Monhagen Hose Company of Middletown, New York, in 1909. This apparatus was built on an American Mors automobile chassis; it boasted a 400 gpm rotary pump.

Numerous firms whose names bring back the heady days when automobiles were novelties also made some fire engines in the first two decades of the 20th century. Kissel, White, Pope, Reo, and others are among them.

The dominant names in the first 20 years or so of the motorized fire apparatus business, however, remained the names that had become famous through their steam engines: American LaFrance, Ahrens-Fox, and Seagrave. These firms were innovative and adaptable, able to cope with a pace so rapid that by 1916 the production of steam-powered fire apparatus had virtually ceased in favor of the automobile.

By 1907, several of the best-known names in fire apparatus were experimenting with or offering motorized equipment. Seagrave delivered its first three rigs (two hose wagons and a straight chemical car) to the Vancouver Fire Department; all used air-cooled four-cylinder engines. Waterous delivered its second motor apparatus, a 600-gpm rig, to the Alameda, California department. Interestingly, this machine utilized the same four-cylinder engine for both propulsion and pumping. Also in 1907 the Howe Fire Apparatus Company weighed in with its first motor pumper, and a new manufacturer, the Webb Motor Fire Apparatus Company, began offering fire engines.

The final gasp of the self-propelled steam pumper came in 1908, when Amoskeag (now a division of the Manchester Locomotive Works) delivered a 16,000-pound behemoth to the Vancouver Fire Department. Top speed was 12 miles an hour, which compared very unfavorably to the Seagraves already in service in the department.

THE CHRISTIE TRACTOR

As the advantages of gasoline power became clear to every fire chief, they all yearned for shiny new motorized fire engines. However, the steam engines already in their station houses were efficient, incredibly sturdy, and (most importantly) paid for. Although there was little justification for replacing the steamers themselves, the horses that drew them were another matter after 1911. In that year, a former racing driver named John Walter Christie came up with a two-wheeled, four-cylinder tractor designed to replace the horses that drew the steamers and other apparatus. His firm, called Christie's Front Drive Auto Company and based in Hoboken, New Jersey, produced over 600 tractors over the next eight years. These powerful little machines had the engine mounted transversely in front, and were capable of pulling 12 tons. Other manufacturers, including American LaFrance, Seagrave, and the C.J. Cross Front-Drive

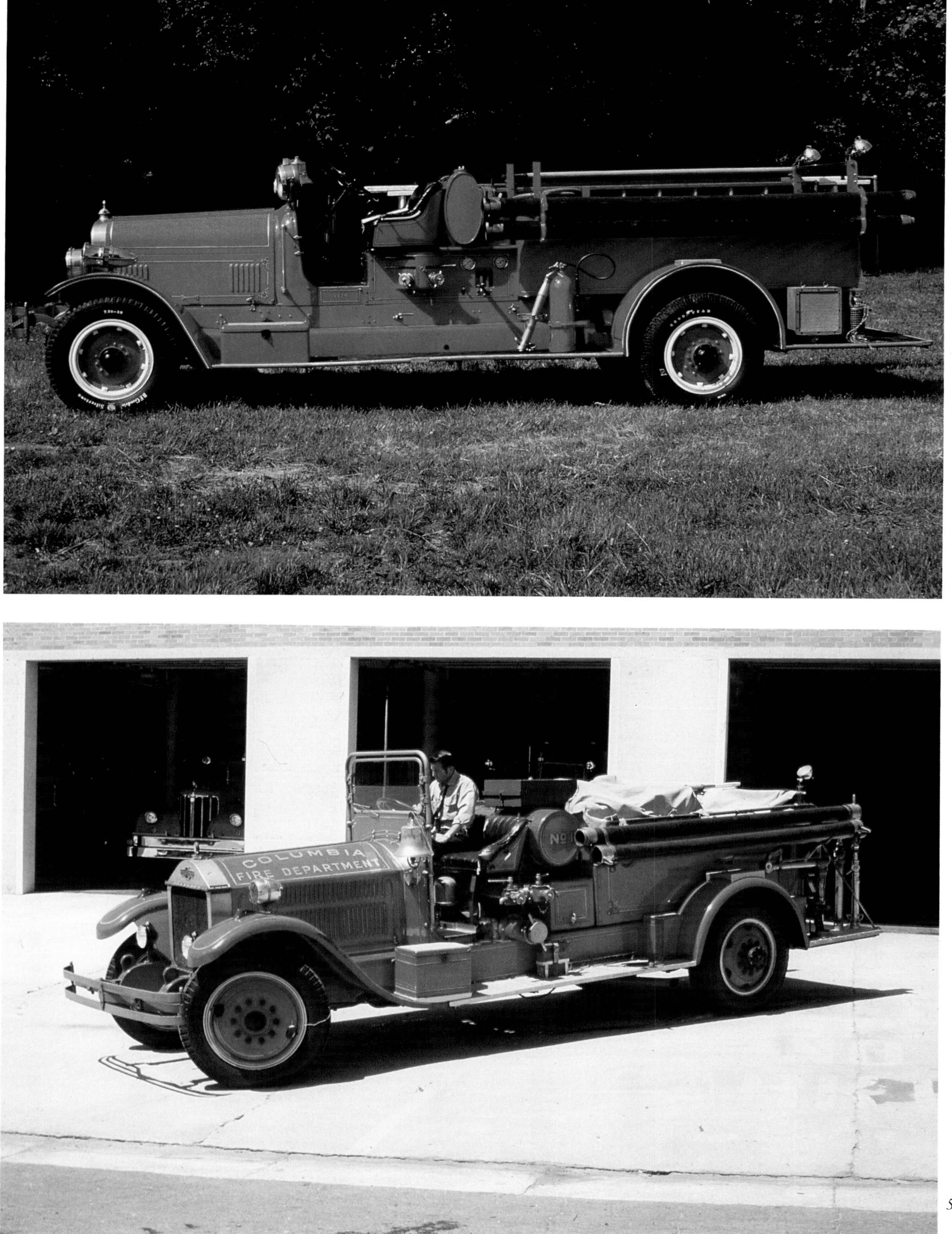

5-51

5-52

5-7

FIRE PATROL No 5
FIRE PATROL
48851

5-8

OSHKOSH MOTOR TRUCK MFG. CO.
"4 Wheel Drive Trucks"

5-9

Tractor Company of Newark, New Jersey (owned by a former employee of Christie) were quick to start making tractors. Christie tractors maintained their lead, however, and were an important aspect of the transitional period from horsedrawn to motorized equipment. By 1912 the New York City Fire Department was operating 28 Christies; by 1917 153 were in service, and the final roster topped 300. The Front Drive Auto Company closed down shortly after the end of World War I, and John Walter Christie's interests turned to another area: armored tank design. Tanks had been introduced during World War I, but they were clumsy, slow, and subject to frequent breakdown in the field. Christie was convinced that tanks were the future of warfare, and he designed a new suspension that was highly effective. With the war over, however, there was little interest in America, and Christie ended up selling only three of his advanced medium tank, called the T3, to the US Army. Interestingly, Christie also sold two tanks to the Soviet Union. These became the model for the famous T34 Soviet tank, which was instrumental in the defeat of the Nazis on the eastern front in World War II.

AHRENS-FOX RETURNS

The birth of one of the most famous names in fire apparatus took place in 1908, when the Ahrens-Fox Fire Engine Company was formed. For the next three years the firm would make only steam engines, particularly the well-regarded Continental model. In 1911 Ahrens-Fox introduced a front-mounted piston pump, powered directly from the crankshaft, that featured two side-by-side domed air chambers; it could pump up to 700 gpm. The first of this popular model, called the Continental Model A, was delivered to Rockford, Illinois in 1912. Ahrens-Fox in 1912 stood in the center of a crossroads, offering several differing types of fire apparatus: traditional steam-powered, horsedrawn fire engines; modern gasoline-powered pumpers; and steam pumpers drawn by gasoline- or electric-propelled motors.

In 1913, Charles H. Fox had yet another inspired idea, one that again changed the face of firefighting. He mounted a small, gasoline-powered centrifugal pump in front of the radiator of a light-duty commercial truck chassis. The pump was fed by an auxiliary water tank located behind the driver's seat. The system quickly became known as the booster tank, and rapidly took over the function of the chemical wagon. It was also quickly incorporated into standard pumpers, thus producing one of the most durable and effective concepts in modern firefighting: the triple combination fire engine, featuring a booster pump, volume pump, and hose.

Ahrens-Fox first demonstrated the booster system at a fire chief's convention in 1913, and soon after delivered ten to the Cincinnati fire department. This was the beginning of the end for the chemical engine, although Ahrens-Fox continued to build chemical engines until 1927, and the very last few chemical trucks were built in the early 1930s. Cincinnati, the home town of Ahrens-Fox,

5-7
Engine Company No. 39 in New York City used this Nott motorized pumper at a three-alarm fire in 1913. The first motor pumping engine in New York City was a Nott that was fitted to a steam engine in 1911.

5-8
NYFD Fire Patrol No. 5 got around in this rig, probably a Webb from around 1912.

5-9
The first plant of the Oshkosh Motor Truck Manufacturing Company, in a photo from 1917.

5-53

5-53
A 1924 Ahrens-Fox pumper on display in Wheaton, Illinois. This view nicely shows the complexity of the piston pump.

5-54
A 1924 American LaFrance Type 75 600-gpm pumper. The rear fender uses a step design.

5-55
This little triple combination is a 1927 American LaFrance run by the Sumpterville, South Carolina, Fire Department. The windshield is a modern addition.

was a loyal customer. By 1918, the department was running nearly 50 pieces of Ahrens-Fox apparatus.

The famous Ahrens-Fox front-mounted pump, with its distinctive large, spherical air chamber, was introduced in 1914 as standard equipment on the firm's new Model K piston pumpers. This rig replaced the Continental twin-dome pumper, which was phased out by 1915. For the next several generations of fire buffs, the large, shiny dome of the Ahrens-Fox pump would be the embodiment of the fire engine.

The next major development at Ahrens-Fox was the Model J, introduced in 1919. This six-cylinder model featured a shaft drive (instead of the chain drive on earlier models) and had a stylishly gabled hood. It was equipped with a 750-gpm pump and would remain in production for the next 11 years.

The famous Ahrens-Fox air chamber went from one-piece to two-piece construction in 1923. The two halves were held together with a band around the equator of the dome, a sure way to identify post-1922 rigs. The first city service ladder trucks from Ahrens-Fox were sold in 1923. Instead of stacking the ladders in a tall and unstable single bank, the ladders were stacked side by side in a double bank. This not only lowered the center of gravity and made the vehicle safer, it allowed more ladders to be carried.

Quadruple combination rigs were first made by Ahrens-Fox in 1927, but they were never particularly popular. Fewer than 30 were sold. New York City in 1927 took delivery of an Ahrens-Fox high-pressure piston pumper that was

5-54

5-55

CHIEF
COPYRIGHT 1910 by Theo. L. LaFrance
ENGINE CO. 5.
H. F. D
COMB'N. HOOK & LADD
H. F. D.
COPYRIGHT 1910 by Theo. L. LaFrance

5-10

5-11

claimed to be the most powerful pumper in the world. In its acceptance tests it threw a vertical stream 800 feet, reaching the top of the first true skyscraper, the Woolworth Building. The Skirmisher line was introduced in 1929. These small quads featured midships-mounted rotary gear pumps, a departure from the usual Ahrens-Fox front-mounted piston pump. A similar departure came in 1930 with the Model V, a small, 500-gpm pumper that proved quite popular. By 1933 Ahrens-Fox had developed a sort of split personality. The company was making extra-large piston pumpers using its traditional front-mounted pumps, but was also planning new, small, inexpensive centrifugal pumpers with midships pumps. A new series of the smaller sizes was introduced in 1935, and another—the SC line—in 1936. The ruinously expensive development costs of the centrifugal rigs, combined with general effects of the Great Depression, put Ahrens-Fox into serious financial difficulties from which the firm never really recovered. In 1939 the firm was forced into liquidation, but it was immediately reorganized and carried on more or less as usual.

The magnificent HT series of piston pumpers began production in 1937. In all, 67 of these 1,000-gpm pumpers were delivered, the last in 1952. To any fire engine buff whose childhood memories encompass this period, the HT will forever be what the words "fire engine" mean.

5-10
A sleek-looking, four-seater chief's car on a Simplex chassis from American LaFrance in 1910.

5-11
The earliest motorized apparatus from American LaFrance was built on the four-cylinder Simplex chassis, as shown here in this photo from 1910.

AMERICAN LAFRANCE INNOVATES

Although American LaFrance had not yet formally entered the automotive field, in 1907 the firm built a straight chemical car on a Packard chassis. This 30-horsepower rig was delivered to Boston, but was returned several months later when the department didn't pay for it. In 1909 American LaFrance delivered two motorized hose and chemical wagons, both built on four-cylinder Simplex automobile chassis. The formal entry of American LaFrance into the motorized apparatus market came in August of 1910, with the delivery of Register No. 1, a Type 5 combination chemical and hose car, to Lenox, Massachusetts. In 1911 American LaFrance began offering rotary-gear pumping engines in two sizes: Type 10 (500 gpm), and Type 12 (750 gpm). Although ALF would begin offering a centrifugal pump in 1916 and piston pumps soon after, for years to come the rotary-gear pump was the company's preference. This was largely a practical matter. Positive displacement pumps were well-suited to the slow speed but high-torque engines of the time. ALF was now using a proprietary chassis for these pumpers; 38 were in service by the middle of 1911. The Type 10 design was also used for hose cars and chemical cars.

In 1914 American LaFrance reached a milestone: the company produced its last steam pumper. (Production of hand pumpers had ended in 1910.) This unit was delivered complete with the new ALF Type 31 tractor to pull it. These front-drive tractors had four-cylinder, 75-horsepower engines; top speed was a mere 25 miles an hour. Nonetheless, the tractors sold well to departments that wanted to motorize their dependable old steam engines.

The smaller Type 40 pumper was introduced in 1915; 250-gpm and 350 gpm versions were available. In 1916 the popular Type 14 model was introduced. Many went into service as combinations or as straight ladder trucks. In 1921 ALF began offering Type 14 in a quadruple combination carrying a full set of ladders, hose, a volume pump, and either a booster pump or chemical equipment. The popular Type 75 standard pumper was being offered by 1922; the favored configuration was the triple combination. The Type 75 powerplant was a six-cylinder, 105 horsepower engine with either worm or shaft drive and a 750 gpm pump. A smaller, light-duty apparatus called the Cosmopolitan was introduced in 1923.

American LaFrance presented a major design change in 1926. The new models were designated Series 100. Perhaps the most popular in the new line was the Type 145 triple-combination Metropolitan.

Another American LaFrance merger took place in 1927, when the firm joined with the Foamite-Childs Corporation (itself formed in 1922 by a merger of several fire-extinguisher manufacturers) to form the American LaFrance Foamite Corporation. American LaFrance achieved a significant breakthrough in 1931, when the firm introduced its first V-12 engines. This massive powerplant, built by ALF, consisted of two six-cylinder engines sharing a single crankshaft. This 240-horsepower engine would power American LaFrance equipment for the next 30 years.

5-56
For firefighters, even riding in a parade can be dangerous work. These men are aboard a 1929 Type 75 American LaFrance.

5-57
The booster hose reel is empty on this 1920 American LaFrance 1,000-gpm combination. The artillery-style spoked wheels have modern tires.

5-67
A close-up look at the piston pump on a 1924 Ahrens-Fox pumper.

In the depths of the Depression in 1937, ALF produced some extremely interesting pumpers for the city of Los Angeles. Both were duplex pumpers, mounting two 1,000-gpm pumps apiece, but the hose was carried in separate manifold wagons. The idea was to reduce the number of rigs near the fire ground, since each duplex pumper delivered the water of three standard pumpers. Los Angeles took delivery of two more duplex pumpers, these having 3,000-gpm capacity, the next year.

American LaFrance introduced the Series 200, also called the Master Series, in 1929; the Series 300 models were first offered in 1933. The beautiful Series 400 design, featuring shaft drive, was introduced in 1935. The justly famous Series 500 models were introduced in 1938. Streamlined and sleek, these wide-body rigs were offered with an optional enclosed overhead ladder compartment that was faired to the cab roof. The compartment added appreciably to the streamlined look; both open and closed cabs were available. By comparison, every other fire engine on the market suddenly looked dowdy.

5-56

5-57

COMBINATION CHEMICAL
H.F.D.
COMBINATION CHEMICAL
HOQUIAM Wn. Fire Dept
©

5-12

5-13

H.F.D.

THE SEAGRAVE COMPANY

The Seagrave Company began selling its Model AC apparatus in 1910. These units were combination hose and chemical wagons featuring the company's characteristic air-cooled "buckboard" style (resembling a buckboard wagon), with chain drive and the driver's seat ahead of the engine. The buckboards continued in successful production until 1915, but the company's first pumper was introduced in 1911. A massive affair with a three-stage centrifugal pump using 11-inch impellers, this rig had a capacity of 1,000 gpm. Under the conventional hood was a huge six-cylinder engine: 9-inch stroke and 7 1/2-inch bore. The pump was enclosed in the rear under a smaller, sloping hood, which gave the rig a rather rakish air. By 1912 Seagrave was offering an entirely new line, with centrifugal pumps ranging from 600 to 1,000 gpm; the engine was under a distinctive gabled hood. It was not until 1921 that Seagrave equipment really changed its looks with the introduction of a rounded hood and radiator.

In 1900 Seagrave started a Canadian subsidiary in Windsor, Ontario. Called the W.E. Seagrave Fire Apparatus Company, it sold the Windsor Fire Department its first piece of motorized apparatus in 1914.

In 1922 Seagrave introduced its first shaft-drive pumper, although it continued to offer chain-drive vehicles for several more years. A significant development came in 1923, when Seagrave introduced the small Suburbanite. Powered by a six-cylinder engine, this popular model had a 350-gpm centrifugal pump. It was an interesting contrast to the Metropolite, Seagrave's largest pumper, which could pump 1,300 gpm. In 1926 Seagrave introduced an in-between size, a 600-gpm centrifugal pumper called the Special.

The Suburbanite model proved to be durable and popular. The New York City Fire Department ordered 50 of them, outfitted with hose and turret, in 1929; these remained in service for many years. The Sentry Series models were added in 1931; a major improvement was the addition of windshields as standard equipment. Another addition was the placement of the siren between the headlights in front of the radiator grille.

Seagrave began offering its first rigs with a V-12 powerplant in 1935. Made by the company, this engine generated 240 horsepower and was ample competition to the American LaFrance V-12. In the next year, Seagrave introduced a new, streamlined look to match the new engine. The distinctive look of the sloping, vertical bars of the slightly rounded radiator grille lasted until 1951, and makes Seagraves from that period readily identifiable.

An important development from Seagrave was the enclosed canopy cab, introduced in 1937. This style remained in production for more than 30 years. It featured an enclosed cab for the driver and officer. The roof of the cab extended back and covered a rear-facing seat holding three or four firefighters. Although many fire chiefs felt that the enclosed design pampered the firemen, the improved safety and comfort of the closed cab soon became apparent. Even so, it would be years before closed cabs were routine and open cabs were found only on antiques.

5-12
The American LaFrance Type 20 combination chemical wagon featured two 35-gallon tanks. This rig ran in Hoquiam, Washington, around 1913. Note the spotlight and hand-cranked siren on the dash.

5-13
This little rig ran in Enid, Oklahoma, in the years before World War I. A chemical tank can be glimpsed in front of the hose basket.

MACK ENTERS THE MARKET

The Mack Brothers Company had its origins in Brooklyn, New York, in 1889. In that year, Augustus Mack, the youngest of five Mack brothers, became a clerk for the Christian Fellesen Wagon Factory. In 1890, his brother Jack joined the company as a stationary steam engineer. In 1893, when Mr. Fellesen retired, the two brothers took over the company; in 1894, their eldest brother, William, joined the firm. Business was not particularly good, however, and the brothers at one point were building milk wagons. The turning point came in 1901, when Jack went for his first ride in an automobile. The driver, a man named Theodore Heilbron, mentioned to Jack that a gasoline motor could be combined with a truck or even a bus. This casual suggestion had awesome repercussions. The Mack Brothers Company built its first bus in the winter of 1902-03, and a second in 1904. In 1904 the company also started producing its own engines.

By 1904 the firm had outgrown its Brooklyn headquarters and moved to Allentown, Pennsylvania. The first motor truck for commercial use was produced at the new factory in 1905. Mack's entry into the fire engine business didn't officially come until 1911, when the company mounted a pump on one of its Senior Series heavy-duty truck chassis and sold it to the Union Fire Association of Lower Merion, a suburb of Philadelphia. In 1914 the small AB model chassis was introduced. The more famous four-cylinder Mack Model AC Bulldog, with the famed sloping nose, was introduced in 1915. Famed for its durability (hence the expression "built like a Mack truck"), the AC Bulldog would have a long history in the fire service. Numerous Mack ACs were built for the Army during World War I. When they later came on the auction block as war surplus, many were snapped up and converted to fire engines. In 1919 the first production line Mack 500 gpm pumper was delivered; the last fire engine built on an AC chassis was sold in 1927. Mack began deliveries of a new line of stylishly designed vehicles, called the Type 19, in 1929. These rigs were powered by six-cylinder, 150-horsepower engines; they were built as triple combinations offering 750- and 1,000-gpm centrifugal pumps. Type 75, 90, and 95 models were also offered in various configurations.

Mack achieved a first in 1935, when it delivered the first completely enclosed sedan-style pumper built to the specifications of the Charlotte, North Carolina Fire Department. This model arguably took the closed-cab concept a little too far. The hard suction hoses were carried on the outside, but the firemen and all their equipment, including the hose, were enclosed in the extended body. However much protection this provided, it simply didn't *look* like a fire engine. Much more in tune with the sensibilities of the time (and of today) was the Mack E Series, an attractively streamlined design offered in 1936. In 1938 Mack began offering the new Type 80, powered by a Mack Thermodyne engine capable of generating 168 horsepower. Available with both closed and open canopy cabs, this model featured a streamlined, sloping grille with a

5-58
A beautifully preserved 1925 American LaFrance on parade in Philadelphia, Pennsylvania.

5-59
The versatile, rugged Model T Ford served widely as the chassis for fire equipment (home-grown and commercial) for many years. This 1924 vehicle was the first motorized apparatus of the Greenwood, Maryland, Volunteer Fire Department.

PHILADELPHIA
SALUTES - I.A.F.C.
BUREAU OF FIRE
1925 AMERICAN
LA FRANCE
PHILADELPHIA FIRE
DEPARTMENT MUSEUM

5-58

Mar-Va
Rentals
CLEANING
GREENWOOD'S
FIRST + OLDEST
PIECE OF EQUIPT

5-59

5-14

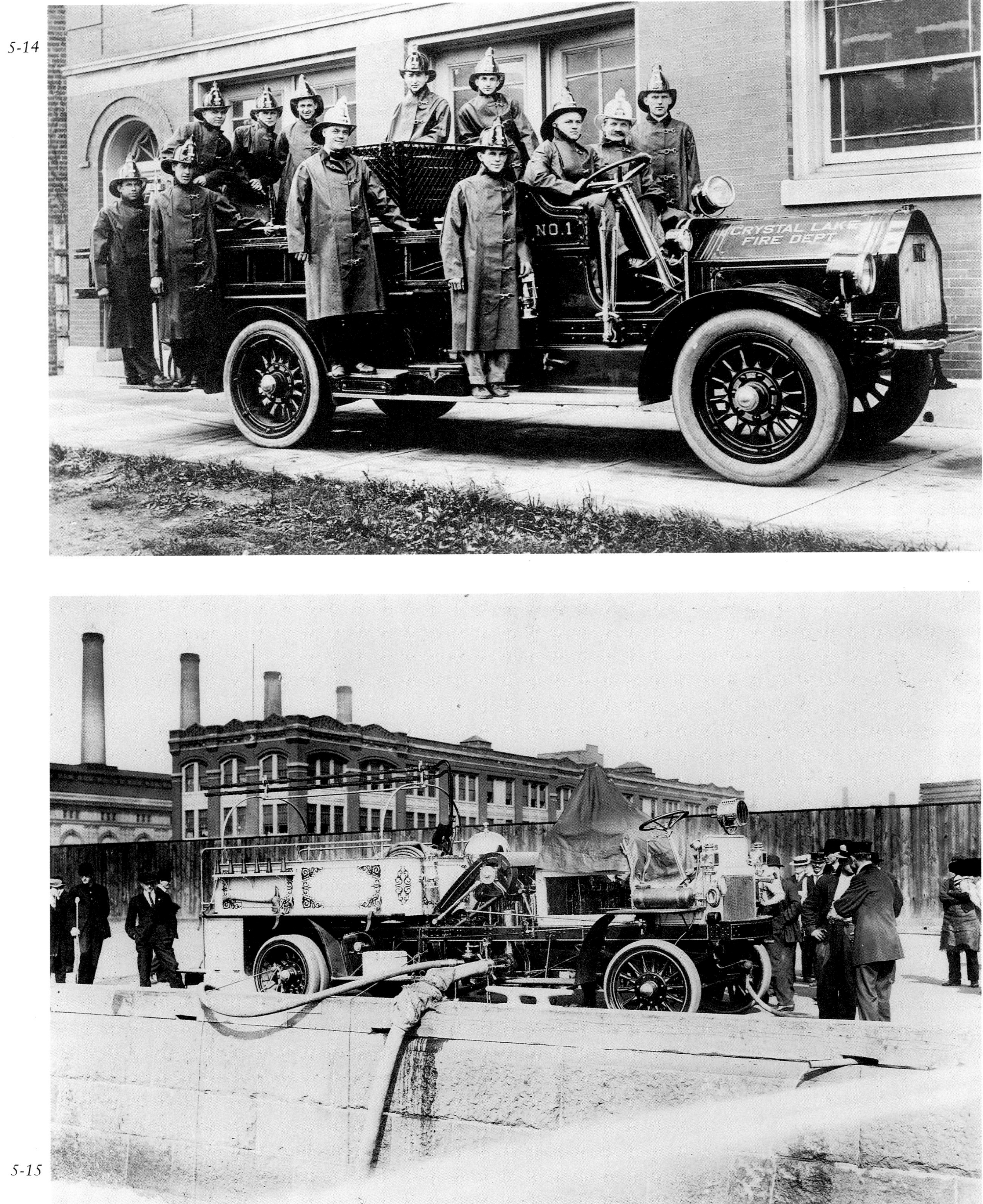

5-15

square mesh and the famous Mack bulldog mascot. The style would remain basically unchanged until 1954.

PETER PIRSCH & SONS

A well-known and long-established manufacturer of ladders and chemical apparatus was the Peter Pirsch & Sons Company of Kenosha, Wisconsin, established in 1857 as the Nicholas Pirsch Wagon and Carriage Plant. Although it had been making hook-and-ladder trucks for years, in 1916 Pirsch delivered its first pumper, a triple combination built on a White Motor Company truck chassis. In 1926 the firm began producing its own chassis. Pirsch apparatus from the late 1920s and into the 1930s carried white hard suction hoses, making them easily distinguishable today (assuming the rig is authentically restored).

Firemen are generally a rugged lot, and apparatus manufacturers saw no need to baby them with such niceties as windshields, let alone enclosed cabs. In 1928, however, Pirsch delivered the first apparatus with a fully closed cab, a 600-gpm pumper that went to Monroe, Wisconsin. This was an early anomaly, however, and closed cabs would not become standard until after World War II. Pirsch delivered closed-cab rigs on a fairly routine basis starting in the late 1930s. Also in the late 1930s, Pirsch began using commercial chassis; GMC models were often employed.

The first Pirsch 1,000-gpm pumper came out in 1929; it used a Waukesha six-cylinder engine producing 130 horsepower. By 1935 Pirsch pumpers were available in every capacity from 500 to 1,250 gallons per minute. A typical rig was the Pirsch Model 20 triple combination. This 500-gpm model was found in many small towns. By 1940 Pirsch apparatus had an attractive streamlined look, often with closed or sedan-style cabs.

5-14
The officers and men of the Crystal Lake FD (state unknown) pose on their combination apparatus in 1914.

5-15
A Seagrave Model AC "buckboard" undergoes acceptance tests at an unknown location in 1911. The cylinder on the fender by the steering wheel is the gas tank.

OTHER MANUFACTURERS

Maxim: The Maxim Motor Company of Middleboro, Massachusetts was founded by Carlton W. Maxim in 1888. In 1914 the firm delivered its first fire apparatus, a motorized hose wagon. Maxim's first motorized pumper was sold a year later. In 1921 the company began offering the M series of pumpers in various capacities. A new B Model was introduced in 1927.

Nott: By 1912 the W.S. Nott Company of Minneapolis, a producer of steam apparatus since 1879, was well established in the motor apparatus field. Nott Universal fire engines were large and surprisingly streamlined in appearance; the transmission was via worm drive. New York City had two 500 gpm Nott triple combination pumpers by 1913; in 1915 the department ordered its first ever combination pumper and hose wagon from Nott. In 1914 a 1,000 gpm Nott Universal was delivered to Victoria, British Columbia.

Four Wheel Drive: Founded in Wisconsin in 1910, the Four Wheel Drive Auto Company (FWD) delivered its first fire apparatus in 1917. True to their

5-60
This 1928 750-gpm Seagrave pumper is painted an unusual and striking black. The louvered hood is open, as it would have been to cool the engine when the pump was in use.

5-61
An American LaFrance triple combination from the 1920s. It is preserved by the Bay City Fire Department of Michigan.

name, these rugged rigs used four-wheel drive and could travel almost anywhere—a feature that made them popular in areas with harsh winters and in areas subject to forest and brush fires. Fire engines were a small part of the firm's operations, but FWD delivered 18 hose and turret wagons to the New York Fire Department in 1928. The FWD chassis was also used by other manufacturers By 1940 FWD had expanded its fire-engine business considerably.

Obenchain-Boyer: The Obenchain-Boyer Company of Logansport, Indiana, never made its own chassis. By 1918 the firm was specializing in converting commercial chassis to fire use. Reo, Ford, and Dodge chassis were often used, but others, such as Sanford and Studebaker, were also employed. Obenchain-Boyer had its heyday in the 1920s, but by the end of the decade was in difficulties. A successor firm, Boyer Fire Apparatus, was established in Logansport, Indiana, in 1929.

Stutz: The dashing Stutz Bearcat sportscar symbolize the Roaring Twenties for many people, but to fire buffs the name Stutz means outstanding fire engines. Harry Stutz entered the apparatus business in Indianapolis in 1919 with a triple combination pumper. The city promptly placed an order for 35 pieces of motorized apparatus with Stutz. By 1923 the firm offered pumpers as large as 1,200 gpm and as small as the 350-gpm Model K, nicknamed the Baby Stutz. The standard engine on a Stutz was an overhead cam built by the company; it produced 175 horsepower. The firm reached its zenith in the mid-1920s, but by 1928, Stutz was out of business. A few rigs bearing the New Stutz name were produced in the '30s, but by 1940 Stutz was gone for good. There was one lasting legacy of New Stutz: the first diesel-powered fire engine. This 1,000-gpm pumper was delivered in 1939 to Columbus, Indiana, where it served for many years.

Buffalo: The Buffalo Fire Extinguishing Manufacturing Company, founded in Buffalo, New York in 1920, specialized in mounting fire engine bodies on commercial chassis, particularly the Larrabee. In the late 1920s, Buffalo began offering its own custom-built chassis, but also continued to offer commercial chassis such as the Ford Model T and Model A. A number of interesting van-style rigs were delivered in the late 1930s. Streamlining arrived in Buffalo in 1939, featuring long hoods and fully enclosed cabs.

Howe: The Howe Fire Apparatus Company, dating back to 1872, produced its first motorized pumper in 1907. Howe usually used commercial chassis, such as the Ford Model A. In addition, in the 1930s the company came out with a new line of custom pumpers called Howe Defenders (a name that is still familiar today), built on the Defiance truck chassis.

Ward LaFrance: It is easy to confuse Ward LaFrance with American LaFrance, particularly since both firms were located in Elmira Heights, New York. The two are today and have always been separate companies. Ward LaFrance was originally a maker of truck chassis, but in the 1930s the firm began offering custom fire apparatus. The first rigs were 500- and 750-gpm pumpers. Acceptance came quickly. In 1938, for example, nine Ward LaFrance

5-60

5-61

5-16

5-17

combination hose and turret wagons were in service in New York City.

Sanford: A small but respected apparatus builder was the Sanford Fire Apparatus Company of Syracuse, in the industrial heartland of New York state. The firm was founded in 1908 as a truck manufacturer, and remained in that end of the business until the mid-1930s. One Sanford truck model of the early 1920s, the six-cylinder Greyhound, became a popular chassis for fire engines. In 1925 Sanford entered the apparatus business directly, offering fire engines built on its Model 500 chassis and equipped with Waterous pumps. The firm also continued to assemble rigs on commercial chassis such as the ever-popular Ford TT. Additional custom models were added in the 1920s, including the 500 gpm Model 528 starting in 1928, and the Cub series, introduced in 1929. The Cubs were available in several different configurations designed for small town use, and they were competitively priced. (A 500-gpm custom pumper from Sanford cost about $6,000 in 1930, as compared to about $7,000 for an equivalent rig from American LaFrance.) This dual strategy gave the company fair success in selling to smaller municipalities in the Northeast.

Sanford struggled through the Depression years. The Cub series was discontinued in 1933 in favor of the new, more streamlined Model 303. Sales were considerably less than rapid, however, and by 1940 the firm was in serious difficulties. The outbreak of World War II brought government orders for apparatus, and the company was able to recover financially and look forward to the post-war years.

THE COMMERCIAL CHASSIS

In general, in the early years fire apparatus makers preferred to use the heavy chassis of dependable luxury automobiles for their equipment. As a result, many early fire rigs were built on chassis from companies whose names are redolent of automotive history: Packard, Locomobile, Simplex, Nash, Pope, and others. A cheaper alternative, and one that was equally reliable if nowhere near as luxurious, was Henry Ford's Model T, introduced in 1908. Over the coming years, this rugged and ubiquitous vehicle would be adapted to fill almost every fire service need, particularly in rural villages and small towns. American LaFrance, for example, built hundreds of small straight and combination chemical cars on Model T chassis. During World War I, the Howe Fire Apparatus Company built 100 fire engines in various configurations for the Army, all on Model T chassis. The introduction of the portable Barton front-mounted centrifugal pump in 1924 made the Model T option even more attractive. The Barton pump took its power from the crankshaft of almost any automobile engine. Weighing only 20 pounds but capable of pumping up to 250 gpm at 20 psi, and capable of lifting water 26 feet, this pump made better fire protection conveniently available at a very reasonable price.

Ford introduced the Model A in 1929. As a chassis for fire engines, the Model A proved as popular as the Model T. When Ford introduced the power-

5-16
The members of Hook and Ladder Company No. 104 of the New York Fire Department pose for the camera. The aerial ladder is drawn by a Christie front-wheel-drive tractor. A gaslight streetlamp is on the pavement at left.

5-17
The life of a fireman is never easy. This ice-covered Seagrave buckboard is at a fire in New York City around 1915.

ful V-8 engine in 1932, numerous manufacturers began using this versatile and economical new chassis. For example, in 1934 the W.S. Darley Company offered a 500-gpm front-mounted pump on a Ford V-8 chassis, complete with everything but hose, for just under $1,500.

The White Motor Company of Cleveland briefly sold its own fire apparatus in the 1920s, but is better remembered for providing a versatile chassis to other manufacturers.

Chassis from General Motors began to be used for fire engines by the early 1920s. Pirsch, for example, used the GMC 2- or 3-man cab chassis for many years. International Harvester also began promoting their chassis around this time, but by 1926 was also producing custom fire apparatus. The C Series was popular in the mid- to late 1930s.

AERIAL LADDERS AND WATER TOWERS

The first motorized aerial ladders came along in 1912, when the New York City Fire Department purchased four aerials from the Webb Motor Company of St. Louis. These interesting experiments were actually a combination of gasoline and electric motive power. A gasoline generator mounted amidships powered individual Couple Gear motors located on each wheel. These cumbersome machines were slower than horse-drawn apparatus, and were quickly superseded by tractor-drawn rigs. Eventually all the major motorized apparatus manufacturers produced aerial ladders and, to a lesser extent, water towers. By the 1930s, metal aerials were becoming available. Stronger, more stable, and sturdier than wooden ladders (and fireproof), metal ladders could be fitted with ladder pipes and thus do the work of the water tower with more flexibility. As the aerial ladder grew in popularity, the water tower gradually faded from the scene and ceased altogether by 1938.

The Vancouver Fire Department in 1909 purchased the first aerial ladder built by Seagrave. This 75-foot model was pulled by a four-cylinder Model AC tractor. By 1914 Seagrave offered a standard, straight-frame 85-foot ladder truck. Seagrave can also claim the distinction of introducing the all-steel ladder in 1935.

Seagrave delivered its first motorized water tower in 1917 to St. Paul, Minnesota; it featured chain drive and a telescoping, 65-foot tower. Amazingly, this rig remained in operation into the 1970s. In 1922 Richmond, Virginia received a Seagrave water tower built on the Suburbanite chassis. An important innovation on this rig was that the spring-raise mechanism was mounted on a turntable, a feature that was previously available only on water towers from Champion. Seagrave's last water tower was a 65-footer delivered to Washington, DC in 1932.

By 1916 the American LaFrance Type 31 straight-frame, spring-raised aerial ladder was in widespread use. The Type 17 tractor-drawn aerial was also pop-

5-62
A very nice Seagrave triple combination from 1929 is shown here. Note the suction intake under the front bumper, the unusual assortment of warning gear on the hood, and the homemade windshield.

5-63
The chain drive can be seen on the rear wheels of this 1923 Model AC Mack Bulldog triple combination.

5-62

5-63

5-18

5-19

ular. Both were available with ladders ranging from 55 to 85 feet. The longest aerial ever made by ALF was a 125-foot Series 500 model, sold to Boston in 1941 (and demolished by a falling wall the next year).

The first factory-motorized water tower was built by American LaFrance and delivered to New Orleans in 1914. In 1930 ALF delivered a new, 65-foot, tractor-drawn water tower to New York City. Capable of throwing an amazing 8,500 gpm through the main mast and two deck turrets, this water tower was said to be the most powerful in the world. American LaFrance also made the last water tower in America, a 65-footer delivered to Los Angeles in 1938.

The first aerial ladder from Ahrens-Fox was built in 1916 for the New Bedford, Massachusetts Fire Department. The chief of the department was E.F. Dahill. Dahill was a friend of Charles Fox, and seems to have persuaded him to build a prototype aerial ladder to test a new concept. The ladder was raised using Dahill's invention, an air compressor powered from the main transmission. The assembly, which came to be called the Dahill Air Hoist was used to raise the 85-foot wooden ladder (built by Peter Pirsch). Although it worked reasonably well, Ahrens-Fox did not get into the aerial business seriously until 1923. In that year, the first of 73 aerials in 17 years was built. Ahrens-Fox built six-wheel, tractor-drawn aerials almost exclusively. They featured six-cylinder engines and Dahill Air Hoists; a characteristic feature is the spotlight directly behind the driver's seat.

The Pirsch Company was famed for its horse-drawn hook and ladder trucks long before it began manufacturing complete fire apparatus, and ladders remained a mainstay of the company. The firm's famed truss-type ladders of the finest Douglas fir had been patented near the end of the 19th century. Pirsch aerials were raised using the Dahill Air Hoist. In 1931, Pirsch offered completely power-operated aerial ladders, a major step forward. These aerials used power take-offs from the 130-horsepower, six-cylinder engines to raise the ladder hydraulically; extension and rotation were done mechanically. The first all-aluminum, multiple-section ladder was introduced by Pirsch in 1936.

Magirus turntable extension ladders had become quite popular in Europe, and were introduced in America in 1927. They never really became popular, and World War II put a temporary stop to their import.

The first aerial ladder truck from Mack didn't arrive until 1929. A power take-off from the tractor engine was used to raise and lower the ladder, which was available in 65- or 75-foot lengths.

WAR AND CHANGE

The Depression years of the 1930s were hard ones for the fire apparatus industry. In 1928, more than 1,500 new fire engines were delivered nationwide; in 1938, that figure had been nearly halved to about 800. Many municipalities simply could not afford to purchase new equipment, even when it was badly needed. Old rigs were repaired and nursed along in hope of better times to come.

5-18
Why firemen should avoid collapsing walls. This Nott rig has been demolished at a fire in New York City around 1920.

5-19
A demolished Nott fire engine of the NYFD is towed away by a Mack "Bulldog" emergency truck sometime around 1920.

The flames of World War II engulfed America in 1941, and the fire apparatus industry rose successfully to the challenges. Fire protection both at home and abroad was a high wartime priority, and many companies benefitted from military contracts for fire-fighting equipment. The Hale Fire Pump Company, for example, was a leader in the development of high-pressure centrifugal pumps in the 1930s. When war broke out, Hale increased production of pump units from 1,000 in 1939 to more than 30,000 in the years between 1940 and 1945. Many of these trailer-drawn pumps went to fight the flames of the London Blitz.

Despite the priority given fire-fighting, the effects of shortages and rationing were felt everywhere. Production of civilian automobiles came to a virtual halt as the automotive industry shifted to a war footing, and anyone (fire chiefs included) needing a new car simply had to make do with an old one. Because fire engines were considered essential in wartime, the fire apparatus manufacturers were able to continue supplying their civilian customers on a limited basis, while also manufacturing equipment for military use. There were serious restrictions, however, and no municipality could order a fire engine unless it could convince several different government agencies that the purchase was absolutely necessary. Once past this hurdle, the fire department had to accept unadorned apparatus that lacked any of the shiny chrome plating and burnished copper parts so beloved by firemen. Chrome was considered a strategic metal and was reserved strictly for military use; copper was so scarce that in 1943 pennies were made of zinc. Civilians had to make do with plain or painted bumpers, pump caps, and the like (many departments had these painted parts plated after the war). The use of sirens on fire engines was also banned in many places; during the war years, sirens were to warn the populace of imminent air attack. Bells and exhaust whistles were used instead.

The industrial might of America won the war, and soon it would turn again to the battle against flame.

5-64

5-64
Designated Emergency Truck No. 5 of the Hagerstown, Maryland, Fire Department, this 1927 Mack Model AB should really be called a high-pressure truck. It carries two deck turrets.

5-65
The wheels on this 1924 Ahrens-Fox 1,000-gpm pumper have the original hard rubber tires. The air chamber on this rig is made of two pieces with a band around the middle, instead of the one-piece air chamber used by Ahrens-Fox until 1922.

5-66
Purchased in 1924 by the Halesite, New York, Volunteer Fire Department, this Ahrens-Fox pumper has been very well preserved.

5-65

HALESITE
FIRE DEPT.
HALESITE
VOLUNTEER FIRE DEPT
LONG ISLAND NY
1924 AHRENS-FOX
ORIGINAL OWNER

5-66

5-20

5-21

5-22

5-20
This Nott steam engine was the first gasoline-propelled pumping engine ever accepted by the New York Fire Department. Assigned to Engine Company No. 58, it went into service in March of 1911.

5-21
A Christie front-drive tractor is underneath the protective tarpaulin in this NYFD photo, dating from around 1920.

5-22
The front end of a Christie tractor is shown nicely in this undated photo of the New York Fire Department in action.

5-23
An American LaFrance combination gasoline/electric aerial ladder truck in a photo from sometime after 1913. The engine divides the cockpit into two compartments, one for the driver (with right-hand drive) and one for the officer.

5-23

5-68

There are NO SHORT CUTS
when it comes to FIRE PREVENTION
HELP KEEP
OUR STATE
BEAUTIFUL AND FIRE SAFE
FIRE CO.
NO. 1
7399
GEORGETOWN
AMERICAN LA FRANCE
1922

5-69

5-70

5-71

5-68
A 1920 American LaFrance combination pumper and chemical truck.

5-69
A 1922 American LaFrance triple combination. This rig had a four-cylinder engine and pumped 600 gpm.

5-70
Firefighting assemblies were mounted on a wide variety of commercial chassis in the earlier days of motororized apparatus. This rig is a triple combination on a 1920 Reo chassis.

5-71
This beautiful American LaFrance Type 12 combination rig dates from 1920.

5-24
Double bucket seats and double chemical tanks are features of this NYFD apparatus.

5-25
This monster apparatus is apparently a horse-drawn, gasoline-powered pumping rig that was in service with the New York Fire Department around 1905.

5-26
A rear view of a horse-drawn, gasoline-powered pumper serving the NYFD around 1905.

5-27
This NYFD aerial ladder seems to have a Couple Gear tractor to pull it. Couple Gear vehicles featured an electric motor on each wheel. They were famous for being slow and for occasionally running out of power before getting to the fire (or back to the station). This picture dates from around 1914.

5-24

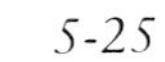

5-25

5-26

5-27

5-72

TEL. 633-9552
OPTIMO
1920
American

1920
American
LaFrance
URBAN

5-73

5-74

5-75

5-72
The American LaFrance Type 12 triple combination could pump 1,000 gpm. This 1920 rig is owned by the Tenafly, New Jersey, Fire Department.

5-73
Veteran apparatus is apt to be cranky. Here members of the Tenafly, New Jersey, Fire Department top up the radiator on their well-preserved 1920 American LaFrance.

5-74
This handsome Seagrave pumper dates from 1928. It is owned by the Ardsley, New York, Fire Department.

5-75
The suction inlet and double discharge gates of a 1928 Seagrave pumper.

5-28
A Christie front-drive tractor has towed this steam pumper to the fire ground in New York City around 1913.

5-29
The full complement of the Lynbrook, New York, Fire Department poses in dress uniform beside a beautiful Mack AC city service ladder truck in the 1920s. A life net is on the running board.

5-30
A 1925 Oshkosh triple combination. The hose reel for the chemical tank is empty. This rig carries 4-inch and 2 1/2-inch suction hoses.

5-31
A Mack model AB fire engine run by the Bethel, Connecticut, Fire Department in the 1920s.

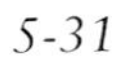
5-31

5-28

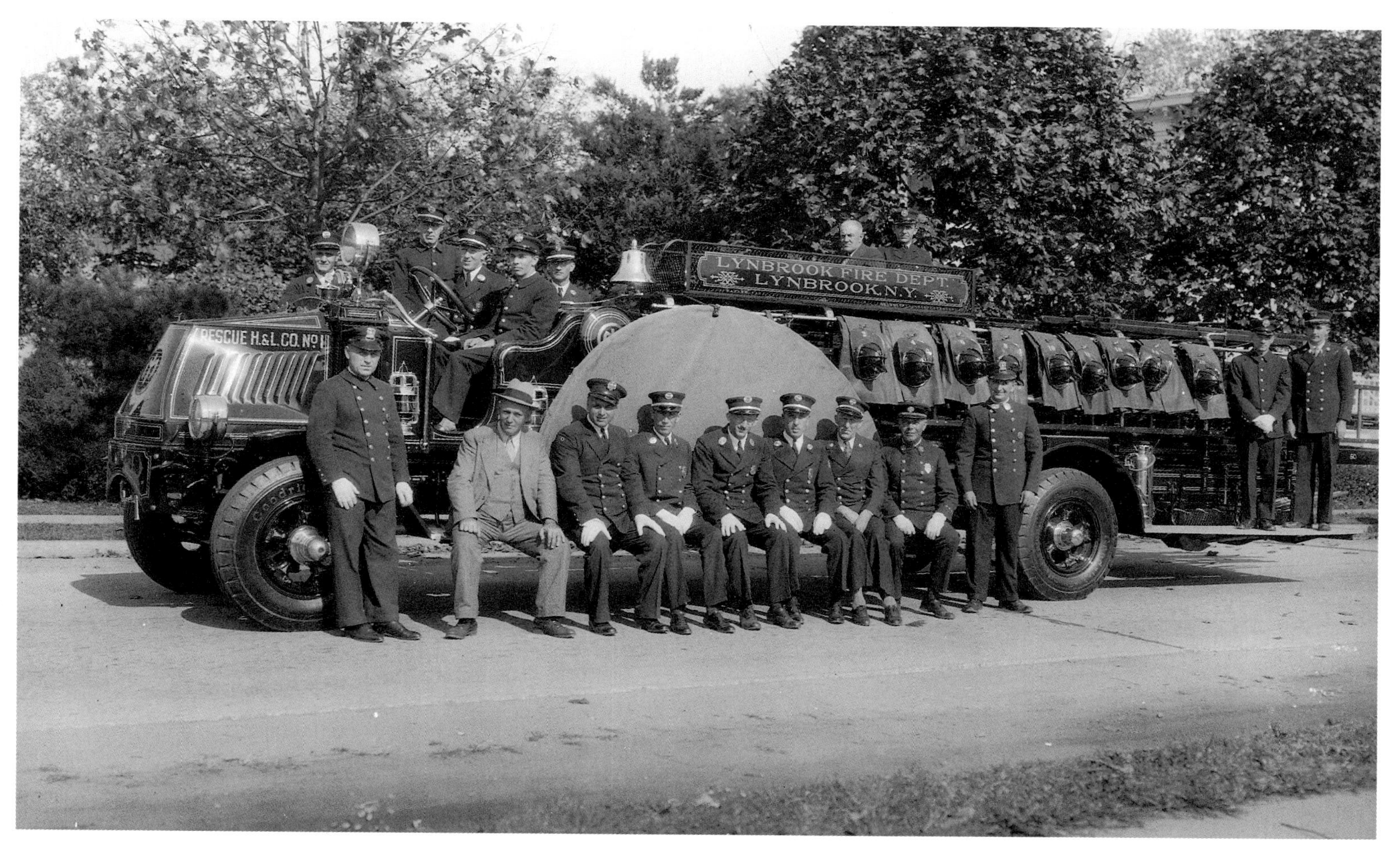
LYNBROOK FIRE DEPT
LYNBROOK N.Y.
RESCUE H.&L. CO. No 1

5-29

5-30

OSHKOSH
753A
O.F.D. No1

5-76

5-77

5-78

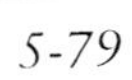

5-79

5-80

5-82

5-81

5-83

5-76
A 1929 Seagrave Suburban pumper, owned by the Beacon Falls, New York, Fire Department.

5-77
A close-up look at the Seagrave badge, on a Suburban from 1929.

5-78
This shot gives a good look at the pump and dash of a 1929 Seagrave Suburban.

5-79
The 2 1/2-inch hose is neatly laid in the back of this 1929 Seagrave Suburban; 1-inch booster hose is coiled in the basket above.

5-80
The traditional fireman's axe rests snugly along the side of a 1921 American LaFrance fire engine.

5-81
The Ford Model TT commercial chassis was the basis for many different fire rigs, particularly in smaller municipalities. American LaFrance built the chemical assembly on this 1924 chassis.

5-82
The Ford logo is easily visible on grille of this Ford/American LaFrance chemical wagon from 1924.

5-83
A stunningly beautiful 1932 Mack Type 70 city service ladder truck. These low-slung, sporty models handled very well.

5-32
Three generations of fire apparatus are displayed at the Oxford, Pennsylvania firehouse in the 1920s. The rig at left probably dates from between 1905 and 1910; the rig at right is an American LaFrance Type 75.

5-33
The city of Indianapolis was protected by some home-grown apparatus. This 600 gpm Stutz was purchased by the city in 1920 and photographed in 1926.

5-34
The men and fire engines of Indianapolis, Indiana Fire Station No. 27. These Stutz engines were photographed in 1926.

5-35
The volunteer fire department of Orange, Virginia, at a Fourth of July parade in 1924. The pumper leading the parade seems to be an Obenchain-Boyer built on a Reo Speedwagon chassis.

5-32

5-33

FIRE STATION NO 27
Indianapolis Fire Department Fire Station #27. 1926.

5-34

29064

5-35

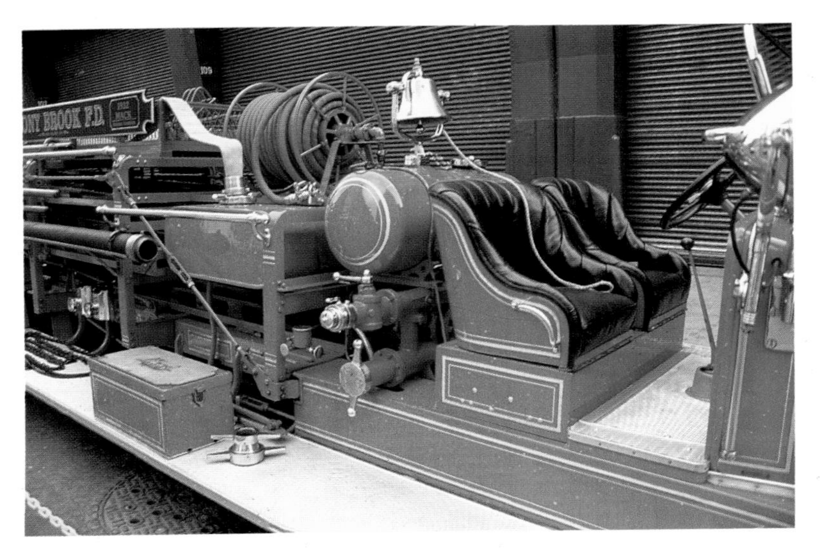

5-84

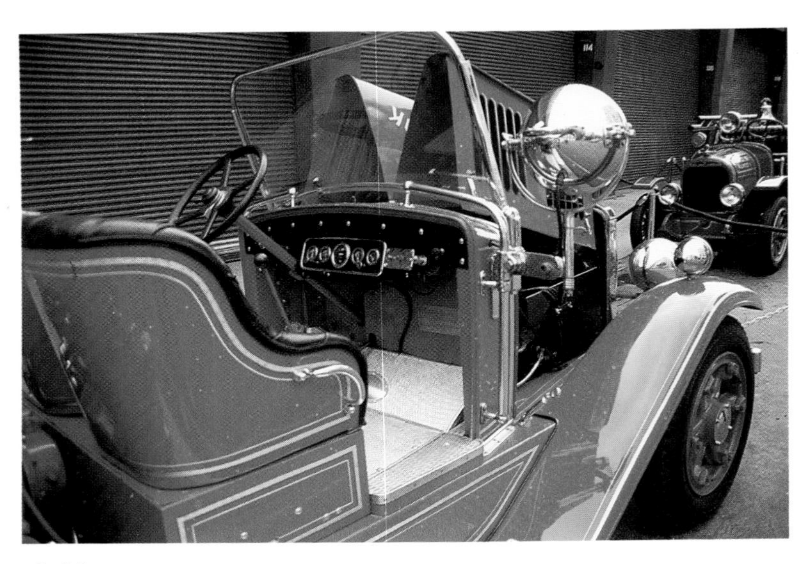

5-85

5-86

5-88

5-87

5-89

5-90

5-91

5-84
This 1932 Mack Type 70 city service ladder truck sported double bucket seats for the driver and officer. This rig is owned by the Stony Brook, New York, Fire Department.

5-85
Louvered hood doors and a simple cockpit distinguish this 1932 Mack Type 70 city service ladder truck.

5-86
Famous around the world, the Mack bulldog mascot glowers from the hood of a 1932 Mack Type 70 ladder truck.

5-87
The pump panel was simpler then. The power take-off lever is at the bottom of this shot, which shows a 1932 Mack Type 70 city service ladder truck.

5-88
This 1932 Mack Type 75 triple combination pumper seems to be missing its hood panels. This rig is run by the Star Hose Company of the Greenport, New York, Fire Department.

5-89
The crew of this 1938 Ward LaFrance rotary pumper poses for the camera at the muster celebrating 125 years of the New York Fire Department, held in Manhattan in the spring of 1990.

5-90
Ward LaFrance began deliveries of this chassis style in 1937. This grille is on a rig from 1938.

5-91
In this close-up look at the front quadrant of a 1938 Ward LaFrance rotary pumper, the post-mounted warning lights are easily seen.

5-39

5-40

5-36
FDNY Searchlight No. 2 leaves fire headquarters on Jay Street in Brooklyn in 1930. This vehicle is built on a 1929 eight-cylinder Packard chassis. The plaque on the door is in memory of Chief Thomas A. Kenny. The cost of this rig was $7,500, a very substantial amount in 1930.

5-38
A Ward LaFrance searchlight truck from 1937. Note the spare tire carried behind the cab and the post-mounted siren in front of the driver's seat.

5-39
Two American LaFrance NYFD rigs from the 1930s.

5-40
A triple combination rig on parade in New York City in the 1930s. Soft suction hose is carried in a doughnut roll on the running board.

BOOKBINDER
SMITH PRINTING CO.
SMITH PRINTING CO.
FIRE HEAD-Q
SEARCHLIGHT 2.
5K 42-03
F.D.N.Y.
SEARCHLIGHT Nº 1.

5-36

5-38

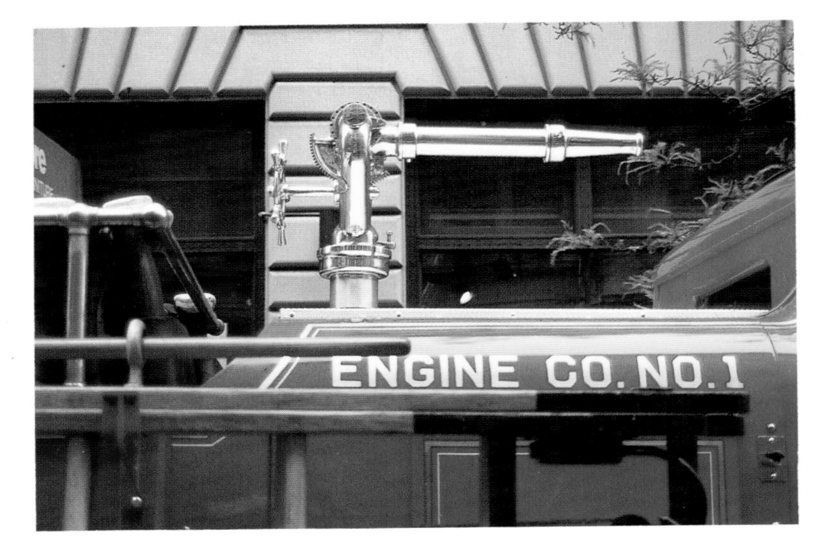

5-92

5-93

5-94

5-95

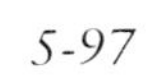

5-96

5-97

5-92
This high-pressure deck gun is mounted on a 1938 Ward LaFrance pumper owned by the Stony Brook, New York, Fire Department.

5-93
The driver's seat and fuel tank of a 1930 Seagrave owned by the Mattituck, New York, Fire Department.

5-94
A good look at the six-cylinder engine on a 1930 Seagrave.

5-95
The badge of American LaFrance as it appeared in 1932.

5-96
This is why Ahrens-Fox pumpers remain for many the ideal of a fire engine. This beautiful, powerful-looking HT rig dates from 1934. It is owned by the Maywood, New Jersey, department.

5-97
The original tires on this 1934 Ahrens-Fox would have been solid rubber on steel-spoked wheels.

5-98
The business end of a 1934 Ahrens-Fox HT pumper. The piston pump put out 1,000 gpm.

5-99
The paintwork and plating on this 1934 Ahrens-Fox HT pumper are beautifully preserved.

5-99

5-41
New York City took delivery of a large order of American LaFrance 750-gpm pumpers in 1921. This group poses on the Grand Concourse in the Bronx after being inspected by Commissioner Doran.

5-42
This Chevrolet tank truck was used for fighting forest fires at the Cleveland National Forest in California. This photo was taken in 1939. The crew accommodations seem to be an old couch.

5-43
A Chevrolet pickup tanker used in Angeles National Forest. It was equipped with a 60-gallon water tank and a Hercules 3/4-inch pump. This shot dates from 1939.

5-44
This 1934 Ahrens-Fox 1,000-gpm pumper served the FDNY into the 1950s.

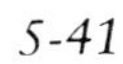

5-41

5-42

5-43

5-44

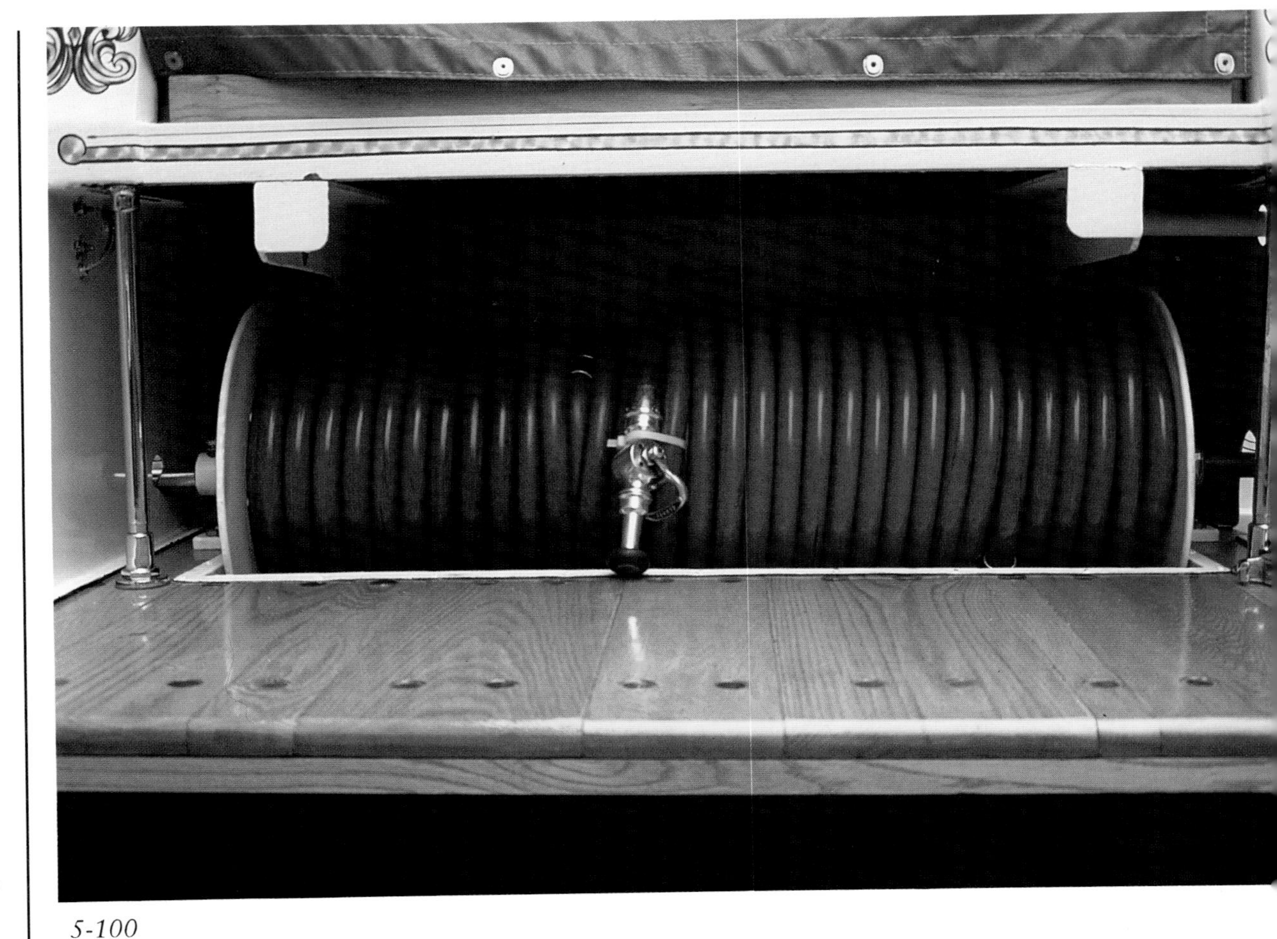

5-100

5-101

5-100
Booster hose is neatly coiled on reels in the back of a 1934 Ahrens-Fox HT combination pumper.

5-101
After the Ford Model T was discontinued, the Model A took its place. This apparatus is a 1931 Ford/Maxim combination pumper owned by the Syosset, New York, Fire Department.

5-102
The booster tank reel rides high above the bed in this 1931 Model A Ford/Maxim fire engine.

5-103
Sturdy Dietz lanterns and warning lights were widely used in the fire service.

5-104
Particularly when compared to modern apparatus, the dashboard of a 1931 Model A Ford seems amazingly simple.

5-105
This electric Federal siren is mounted on the front bumper of a 1931 Model A Ford/Maxim fire engine.

5-106
A head-on look at the grille of a 1940 Ward LaFrance pumper.

5-107
The pumping panel of this 1940 Ward LaFrance rig is simplicity itself, especially compared to today's versions. This pumper, owned by East Brunswick, New Jersey, remained in service until 1971.

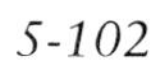

5-102

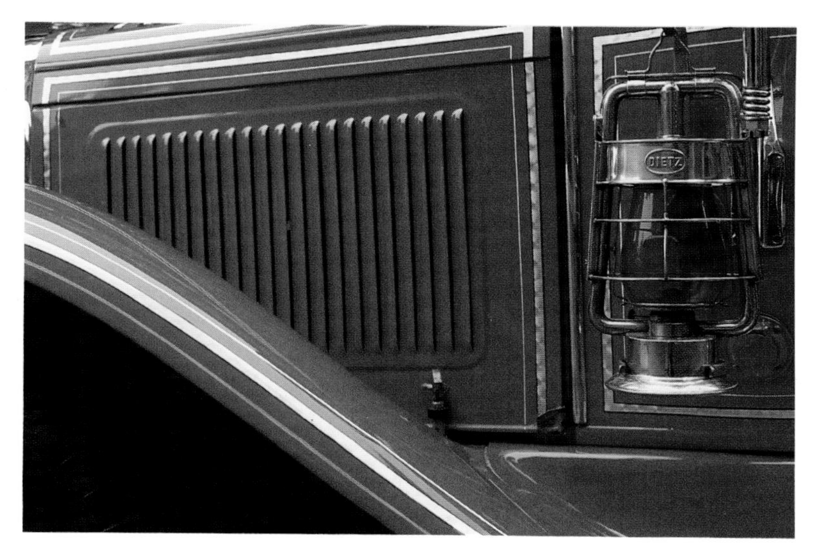

5-103

5-104

5-105

5-106

5-107

5-45

5-46

5-47

5-48

5-45
A Mack Model AC 700-gpm pumper from 1930 at a New York City fire scene.

5-46
The forest-fire suppression team at Sierra National Forest in California got around in a modified Ford pickup. This picture was taken in 1940.

5-47
This 1940 photo shows an International harvest 3/4-ton fire truck at Chippewa National Forest in Minnesota. This rig carried a five-man crew and a 60-gallon water tank. A close look at the area above the truck bed reveals a wire aerial for radio communications.

5-48
A fire truck used to fight forest fires in Montana in a photo from 1936.

5-108

5-109

5-110

5-111

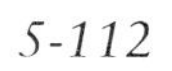

5-112

5-113

5-114

5-115

5-108
White-painted apparatus was not really unusual in the first half of the 20th century. This Ward LaFrance pumper dates from 1940.

5-109
Mack began delivering fire apparatus with this sort of grille in the late 1930s. This is a 1937 Mack pumper owned by Chappaqua, New York.

5-110
A rear view of a 1937 Mack pumper reveals the neatly laid hose, a strainer for the hard suction hose, a spare nozzle, and brass fire extinguishers.

5-111
Chevrolet was a popular chassis for fire assemblies. This little Chevrolet/Howe combination dates from 1934.

5-112
The booster tank and reel can be seen atop this Chevrolet/Howe from 1934.

5-113
A Federal Q siren is mounted on the wooden runningboard of a 1934 Chevrolet/Howe combination rig.

5-114
Spare nozzles and Dietz warning lanterns ride the wooden tailboard of a 1934 Chevrolet/Howe combination. Note the Chevy logo on the end cap at left, next to the lantern.

5-115
By the late 1930s the Sealand Corporation was producing fire engines. The rig shown here was made in 1939 and is owned by the Centereach, New York, Fire Department.

5-138

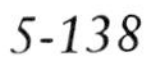

5-139

5-138
The very first motor-driven ladder truck in the United States was built by the International Motor Company of New York City, the predecessor firm to Mack Trucks. It was sold to Allentown, Pennsylvania, in 1909.

5-139
America's first ladder truck with a gasoline engine poses with ladder extended. This shot was taken in Allentown, Pennsylvania, shortly after the rig was delivered by International Motor Company in 1909.

5-140
One of the earliest Mack fire rigs is this city service hook and ladder truck, delivered to Morristown, New Jersey, in 1910.

5-141
Although there had already been some Mack fire apparatus deliveries, this rotary pumper was the first formal sale. It was delivered to the Union Fire Association of Lower Merion (now called Bala Cynwyd), Pennsylvania, in 1911. The big air chamber is made of brass.

5-142
The famed Mack Model AC "Bulldog" was introduced in 1915. This versatile chassis found wide application in the fire service, as shown by this aerial ladder in Hartford, Connecticut.

5-140

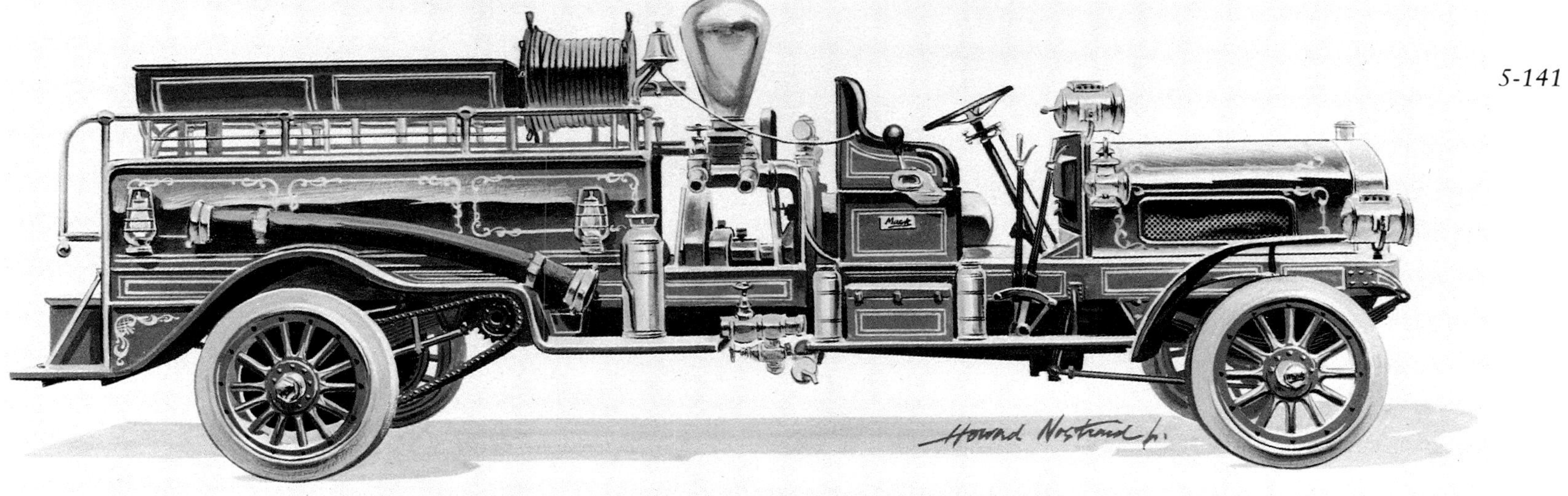

5-141

5-142

5-116

5-117

5-119

5-119A

5-120

5-121

5-122

5-123

5-116
This Sealand fire engine from 1939 is built on an International Harvester commercial chassis. The Sealand Corporation was based in Southport, Connecticut.

5-117
This interesting pumper, built on a 1937 Diamond T commercial chassis, is owned by the Grafton, Ohio, Fire Department.

5-119
A 1941 Seagrave 85-foot tractor-drawn aerial ladder. A close look above the license plate reveals a relic of the Cold War—a Vermont Civil Defense sticker.

5-119A
Calais, Maine owns this 1931 Mack city service ladder truck.

5-120
A 1933 Mack combination pumper, owned by the Dresden, New Jersey, Fire Department. Note the doors instead of louvers on the hood.

5-121
An array of spotlights adorns this 1930 Seagrave Suburban.

5-122
The Sentry series of closed-cab apparatus was introduced by Seagrave in 1937. This pumper, run by the Newark, Ohio, Fire Department, dates from 1939.

5-123
A Barton pump is mounted on the front of this Reo Speedwagon chassis from 1935. Inexpensive, efficient, and reliable, Barton pumps were widely used by rural fire departments.

5-143
Mack introduced the Model 19 triple-combination pumper in 1929. The suction port, visible here between the feet of the two firemen on the running board, is placed below the frame. This unit went into service with the Cincinnati Fire Department in early 1933.

5-144
The Fairview Fire Company of Allentown, Pennsylvania, took delivery of its Mack Type 70 600-gpm centrifugal pumper in April 1930—one of the first deliveries of this type rig.

5-145
A Mack Type 90 aerial ladder undergoes acceptance testing in Montclair, New Jersey, in March 1931.

5-146
The Mack Type 50 was a small fire engine that pumped 500 gpm. This Type 50 quad dates from 1935.

5-143

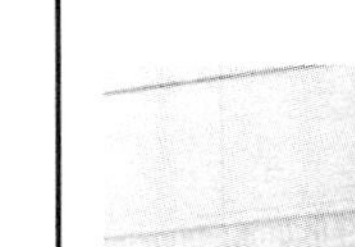

5-144

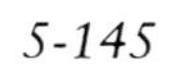
5-145

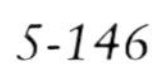
5-146

5-124

5-125

5-124
The Pirsch trademark bell is on the front of this 500-gpm pumper from 1932.

5-125
A snazzy-looking Diamond T chassis was used to build this 1936 Pirsch combination. It is owned today by the North Branch, New Jersey, Fire Department.

5-126
This 1933 Ahrens-Fox pumper was once owned by the Vigilant Fire Company of York, Pennsylvania. It is now on display at the Fire Museum of Maryland.

5-127
Dating from 1940, this Ahrens-Fox pumper demonstrates the correct way to connect the hard suction hose.

5-126

5-127

5-148

5-147

5-149

5-150

5-147
Mack introduced the large Type 95 in 1932. Toronto, Ontario, received this Type 95 aerial ladder in 1934. Note the stanchion-mounted Roto-Ray warning flashers on the running board.

5-148
A Mack Type 95 aerial ladder undergoes predelivery testing in 1934.

5-149
The first sedan-cab pumper in America was built by Mack for the Charlotte, North Carolina, Fire Department in 1935. Known as the Mack Fire Sedan, this completely enclosed model pumped 750 gpm.

5-150
The small Mack Type 55 pumper shown here dates to 1934. It was run by the Lynnfield, Massachusetts Fire Department.

5-128
This 1938 Mack city service hook and ladder truck served Charleston, South Carolina. Note the large life net attached to the runningboard.

5-129
A handsome Ahrens-Fox pumper from 1939 sports a siren mounted directly in front of the spherical air chamber.

5-130
Roto-Ray flashers are mounted on the left-hand side of this Ford/Howe triple combination pumper from the 1930s.

5-131
Retired from active service, this American LaFrance city service truck dates from 1937.

5-132
A modern aluminum ladder has been added to this 1941 Seagrave city service ladder truck from Van Cortland, New York.

5-128

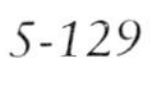

5-129

5-130

BORDER
DEDICATED TO
COMMUNITY
YOUTH
ACTIVITIES
VAN CORT.H.&L.2
HX-3146

5-131

5-132

5-154

5-153

5-152

5-151

5-151
A Mack Type 55 quad rig built in 1937. Note the enclosed overhead compartment for the ground ladders.

5-152
Wartime austerity meant that these Mack Type 80 pumpers were delivered to Fort Worth, Texas, with no chrome trim and—even worse—no Mack bulldog.

5-153
Starting in 1937, the most powerful pumper Mack offered was the Type 21, which could move 1,500 gpm. This photo of an enclosed model used by the Monterey Park, California, Fire Department is dated July 1946.

5-154
This unusual rig is a Mack Type 75 pumper used in Cadillac, Michigan. The canopy cab seats five crew members. A catwalk connecting to the cab is between the two hose beds. This apparatus was delivered in 1941.

5-133

5-134

5-135

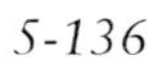

5-136

5-137

5-133
Small American LaFrance combination fire engines like this one from 1933 were the backbone of many fire departments well into the 1950s.

5-134
The distinctive Ward LaFrance grille and a streamlined pumping panel distinguish this 1940 pumper.

5-135
This 1945 American LaFrance Series 500 combination pumps 500 gpm. It is one of the last of this series ever made.

5-136
The Ford Model AA commercial chassis replaced the Model TT in the mid-1920s. This Howe/Ford rig dates from 1928. The three containers on the running board contain water (white), gas (red), and oil (black).

5-137
A classic well deserving of its museum setting is this Ahrens-Fox pumper. It is preserved at the CIGNA Museum and Art Collection in Philadelphia.

6-1

6-2

THE MODERN ERA

The years from World War II to the present have seen tremendous change and many improvements in the fire service. New firefighting concepts, such as the use of foam and high-pressure fog, were developed during the war and have made their way into civilian firefighting. Over the decades since the mid-1940s, new apparatus designs and vast improvements in personal gear have made the job of the firefighter considerably safer, even as that job has changed enormously. Today's fire service does far, far more than simply fight fires.

RADICAL CHANGE AFTER THE WAR

As the fire apparatus manufacturers returned to civilian production in the years after World War II, they were inundated with orders from localities that had not been able to purchase new equipment for years—many since the 1920s. The demand for new equipment was not simply to replace worn-out rigs. As suburban housing tracts began to rapidly replace open farmland, a significant new demand for professional fire protection was created. Fortunately, many returning soldiers and sailors had been trained in firefighting during their service, and some of them would become professional firemen.

Even before the war was over, engineers at American LaFrance were working on a radically different new approach to fire engines: the cab-forward design. This sleek, snub-nosed design, dubbed the Series 700, positioned the cab ahead of the engine and the front axle. The most obvious advantage of the new design was sharply improved forward vision for the driver. The design was also compact and had a smaller turning radius, making it more maneuverable around corners and on crowded city. The cab was available in both open and closed versions. Either one could hold three firefighters in the front seat, with space for two more in jumpseats built into either side of the engine compartment. In addition, three firefighters were carried on the back step. These rigs were powered by the famous American LaFrance V-12 engine.

The Series 700 prototypes were built in 1944, but the model was not officially announced until December of 1945, and the first production models were not delivered until 1947. Production of the Series 500 and 600 models continued through 1947, but American LaFrance had made a major commitment to

6-1
NYFD fireboats pour water on a pier fire in 1947.

6-2
A firefighter mans a huge deck gun on the NYFD fireboat Firefighter *in the late 1940s.*

6-22
Saulsbury custom-built this 18-foot Cardinal series rescue vehicle on a Kenworth 900 chassis. A four-bottle, 5,000-psi air cascade system was installed.

6-23
This Cougar series rescue unit from Saulsbury is installed on an International/Navistar chassis. The six 1,500-watt quartz floodlights are on a 25-foot extendable mast that rotates through 360 degrees.

the Series 700. The factory in Elmira was retooled on the basis of the new design, and by 1947 was producing practically nothing but Series 700 apparatus.

The Series 700 was an instant success. Given the usual stubborn conservatism of the fire service, immediate acceptance of such a radical change may seem surprising, but the obvious advantages of the new design very quickly outweighed any doubts.

The first two Series 700 rigs to be delivered were 1,250-gpm closed-cab models sent to Elkhart, Indiana. A significant breakthrough in 1947 was an order from New York City for 20 open-cab, 750-gpm pumpers with turret pipes. This was the first large order for the new model. Many, many orders from municipalities large and small followed. For any fire buff whose formative years were in the 1950s, the Series 700 (preferably with the open cab) is the sine qua non and ne plus ultra of fire engines.

The depression of the 1930s was hard on many apparatus firms. Most municipalities simply couldn't afford to buy new equipment unless it was absolutely necessary. Many made do with often-repaired and sometimes inadequate equipment. When new rigs were needed, top-of-the-line custom models were avoided in favor of commercial models. This approach affected every fire apparatus manufacturer, but the Ahrens-Fox Company was particularly hard hit. Ahrens-Fox had a reputation for outstanding but pricey quality; its products were sometimes called the Rolls-Royce of fire engines. In addition, the firm's famous piston pumps were complex and expensive to maintain. In 1936, Ahrens-Fox found itself badly undercapitalized. The apparent solution was a merger with the LeBlond-Schacht Truck Company, another Cincinnati firm. One result was that a number of small, rotary-pump fire engines were built on the LeBlond-Schacht chassis. The merger could not save the proud company, however, and in 1939 Ahrens-Fox went into liquidation. Reorganization followed almost immediately, and Ahrens-Fox did manage to deliver its one and only quintuple combination fire engine to Fairview, New York in 1939—a $20,000 monster rig pumping 1,000 gpm and sporting a 70-foot aerial ladder. In the years to come, Ahrens-Fox would deliver a number of EC centrifugal pump rigs, but the famed front-mounted Ahrens-Fox piston pump was rapidly becoming a thing of the past. The last aerial ladder trucks from Ahrens-Fox were delivered in 1940. Peacetime found Ahrens-Fox building pumpers and combination rigs using both piston (Model HT) and centrifugal (Model VC and HC) pumps. The firm's last city service ladder truck was delivered in 1946. In the period between 1946 and 1950, the firm delivered only 89 fire engines, of which 28 were piston pumpers. This was clearly not enough business to keep the company going. In 1950, Ahrens-Fox was once again in financial trouble, and the firm was once again liquidated. This time the assets were purchased by Cleveland Automatic Machine Company and reorganized into the Ahrens-Fox Division. Production of both piston and centrifugal pumpers continued, but fewer and fewer of both models were ordered. The new reorganization lasted for only a year. In 1951, the Ahrens-Fox Division was sold to Gen-

ORLAND
FIRE
PROT. DIST.
KENWORTH
6005

6-22

6-23

OCEANSIDE
NEW YORK
OCEANSIDE F.D.
RESCUE CO.1

6-3

eral Truck Sales Incorporated, also of Cincinnati.

Not every manufacturer that survived the war years went on to success in the postwar period. Buffalo Fire Appliance Corp. went out of business in 1948. A surprising number of Buffalo apparatus survive today as parade pieces in small towns throughout upstate New York.

THE FIFTIES

Considering the state of its business, Ahrens-Fox was lucky to celebrate its centenary in 1952. A more somber milestone was the delivery of the firm's very last piston pumper, a 1,000-gpm model HT for Tarrytown, New York. The new parent company, General Truck Sales, was a distributor for General Motors trucks. Part of the strategy for reviving the firm was for Ahrens-Fox to make 750-gpm centrifugal pumpers on standard GM chassis. It didn't work. Few of these commercial rigs were sold, and by 1953 the firm once again had a new corporate relationship. This time the optimistic buyer was the C.D. Beck Company of Ohio, a firm best known for building buses. Surprisingly, after this change Ahrens-Fox did not quietly slide into oblivion. On the contrary, in 1956 the venerable old firm introduced an outstanding new line of cab-forward models. Configured with both closed and open cabs and with centrifugal pumps rated from 500 to 1,500 gpm, these were at the time the only other cab-forward designs available aside from American LaFrance apparatus. The new models offered excellent engineering, construction, and driver visibility. The closed cab offered seating for five firefighters, and an additional two could be accommodated in seats under a rear-facing canopy.

No sooner had this model been released to universal acclamation than the C.D. Beck Company was purchased outright by Mack Trucks in 1956. Only six of the new design were built as Ahrens-Fox rigs. As will be discussed below, the innovative cab-forward design was quickly redesignated the C Series by Mack. The design would go on to immense success, but the firm of Ahrens-Fox was now defunct.

By 1955 American LaFrance was offering three smaller, less expensive versions of the Series 700 featuring six-cylinder engines manufactured by Continental. The Ranger model had a 500-gpm pump; the Protector had a 750-gpm pump; and the Crusader had a 1,000-gpm pump.

The major plant investment required to retool for cab-forward production had placed American LaFrance in a difficult financial position, despite healthy sales. In order to finance additional growth and ongoing research and development, in 1956 the firm was sold to Sterling Precision Corporation. For ALF clients, the change was inconsequential and business continued as usual.

The Series 700 was replaced in 1956 by the Series 800 models. The new design was basically a refinement of the Series 700, and outwardly there was little difference aside from the addition of equipment compartments built into the rear fenders and the moving of the pump panel from the right to the left side.

6-3
A deck gun on the NYFD fireboat Firefighter. *Note the numerous valves around the base of the gun.*

6-24
A custom-built Concorde series 20-foot rescue vehicle built by Saulsbury. The chassis is a custom Spartan Gladiator tilt-cab model.

6-25
Saulsbury cutom-built this 1,500-gpm pumper/rescue apparatus on a Pemfab four-door tilt-cab chassis.

Internally, the chassis was heavier. More importantly, the parallel-series pump was the new, improved Twinflow model.

The designers at American LaFrance were unsatisfied, however, and the Series 800 was replaced by the Series 900 after only two years. In 1958 the first of many thousands of Series 900 rigs were delivered. The new design featured a wider cab with a wraparound windshield and a choice of eight different engines.

Mack Trucks continued to manufacture fire engines through the war years. In 1941 New York City took delivery of a dozen 1,000-gpm Type 21 pumpers, and added another ten in 1944. When peacetime production resumed in full, Mack was offering several popular models: the Type 45 500-gpm pumper; the Type 75 750-gpm pumper with open cab; the Type 85 750-gpm pumper with closed cab; and the Type 95 1,000-gpm pumper. The Type 19 city service ladder truck was also in production. All the Mack models were quite popular and in great demand. In 1950, for example, the St. Paul, Minnesota Fire Department purchased 14 Type 95 triple combination pumpers; in 1954 Chicago purchased 30 Type 95 double combination pumpers.

In 1954 Mack introduced a completely new line called the B Model. Powered by the Mack Thermodyne six-cylinder engine, this successful line had a distinctively large, plated radiator grille that makes it easy to identify today. Because numerous cab configurations and pump capacities (from 500 to 1,250 gpm) were available, and because the model remained in production until 1967, hundreds of B Models were soon found in firehouses throughout the country. It was on the sturdy B Model that Mack installed one of its Thermodyne diesel engines in 1959. By 1962 diesel power was an option on all Mack apparatus.

When Mack took over Ahrens-Fox in 1956, the cab-forward design was renamed the C Series. This excellent design would stay in production for over a decade. It was quickly off to a flying start when a large order for the new pumpers was placed by New York City in 1957; these rigs had the first automatic transmissions ever used by the NYFD. So pleased was the department by the new apparatus that Mack won the huge contract to upgrade the entire fleet. This meant that over 60 C Series pumping rigs were delivered in 1958, and 20 more were added in 1959. NYFD fans can easily spot these rigs: The old-fashioned curved fenders have no built-in compartments.

The Seagrave Corporation was another firm that emerged from the war years in good shape. In the late 1940s Seagrave introduced a short canopy cab with small, usually oval side windows. This design remained in use on all closed-cab Seagraves until the end of the 1950s, making rigs from this period instantly identifiable. Open-cab models from the period can be identified by the rakishly sloped windshield. (Virtually any Seagrave can be spotted by the bell mounted in front of the right-hand door.) Seagrave produced custom apparatus, but was better known for smaller pumpers produced on commercial chassis.

RESCUE 1
BREWERTON FIRE DIST.
RESCUE 1
BREWERTON
FIRE
DISTRICT

6-24

6-25

SQUAD 14
FIRE CO.
FIRE & RESCUE

6-4
A crow's-nest water tower being raised on the NYFD fireboat Firefighter.

6-5
New York City was well-known for the subway-style hanging handholds used by firemen riding the back step, as shown on this American LaFrance Series 700 pumper from 1947.

A significant anniversary for Seagrave was its seventieth in 1951. To commemorate the event, Seagrave introduced a new line of fire engines called the 70th Anniversary Series. So successful were these rigs that they were still being made 19 years later as the firm neared its ninetieth anniversary. The Anniversary Series rigs were streamlined and modern-looking with built-in equipment boxes on the rear fenders, but they retained many tried-and-true elements. Most importantly, they remained engine-ahead designs with traditional V-12 engines and centrifugal pumps. A characteristic feature is the siren built into the nose of the truck.

Seagrave was not unalterably wedded to the engine-ahead style. In 1959 the firm introduced a popular cab-forward design, still powered by a V-12 engine and with the siren still mounted on the now-flattened nose, that was soon outselling the Anniversary Series. This basic design continues to be offered, with improvements, today.

In 1956 the Maxim Motor Company was wholly acquired by Seagrave. The two companies continued to function as entirely separate units. In 1963, however, Seagrave (and thus also Maxim) was purchased by the Four Wheel Drive Corporation and became known as the Seagrave Fire Apparatus Division. Production was moved from Columbus, Ohio, to the FWD plant in Clintonville, Wisconsin. After the merger Seagrave became best known for aerial ladders and elevating platforms, although fire engines continue to be a significant part of the company's operations.

During the war years Peter Pirsch & Sons continued to produce some streamlined models with enclosed sedan-style cabs, but the firm concentrated on defense work. After the war, Pirsch carried on, building custom rigs using the same six-cylinder Waukesha engine as before. Pirsch also continued its tradition of building apparatus on whatever commercial chassis the customer requested. This means that Pirsch turned out rigs on Chevrolet, GMC, International Harvester, and Ford chassis, among others. For small departments, these rigs are optimal, since they are relatively inexpensive and can be serviced locally.

The sturdy Howe Defenders of the 1930s got a new look in 1953, when the firm began building them on a new, larger chassis powered with a Waukesha six-cylinder engine. These were big rigs, offered only in 750- 1,000-, and 1,250-gpm configurations. In 1960 Howe started offering the Defender on a new, five-man cab-forward chassis made by Truck Cab Manufacturers Inc. of Cincinnati. This canopy cab on this model was so well-designed that the style was soon adopted by virtually every apparatus builder. Hence the style has come to be known as the Cincinnati cab, even when it is manufactured elsewhere. Howe delivered a number of custom Defenders with the new design, but commercial chassis continued to be the firm's chief business.

During the boom times of the immediate postwar period, Ward LaFrance received an order from New York City for 20 750-gpm pumpers. The first was delivered in 1946. Around this time Ward LaFrance began using a distinctively

6-4

6-5

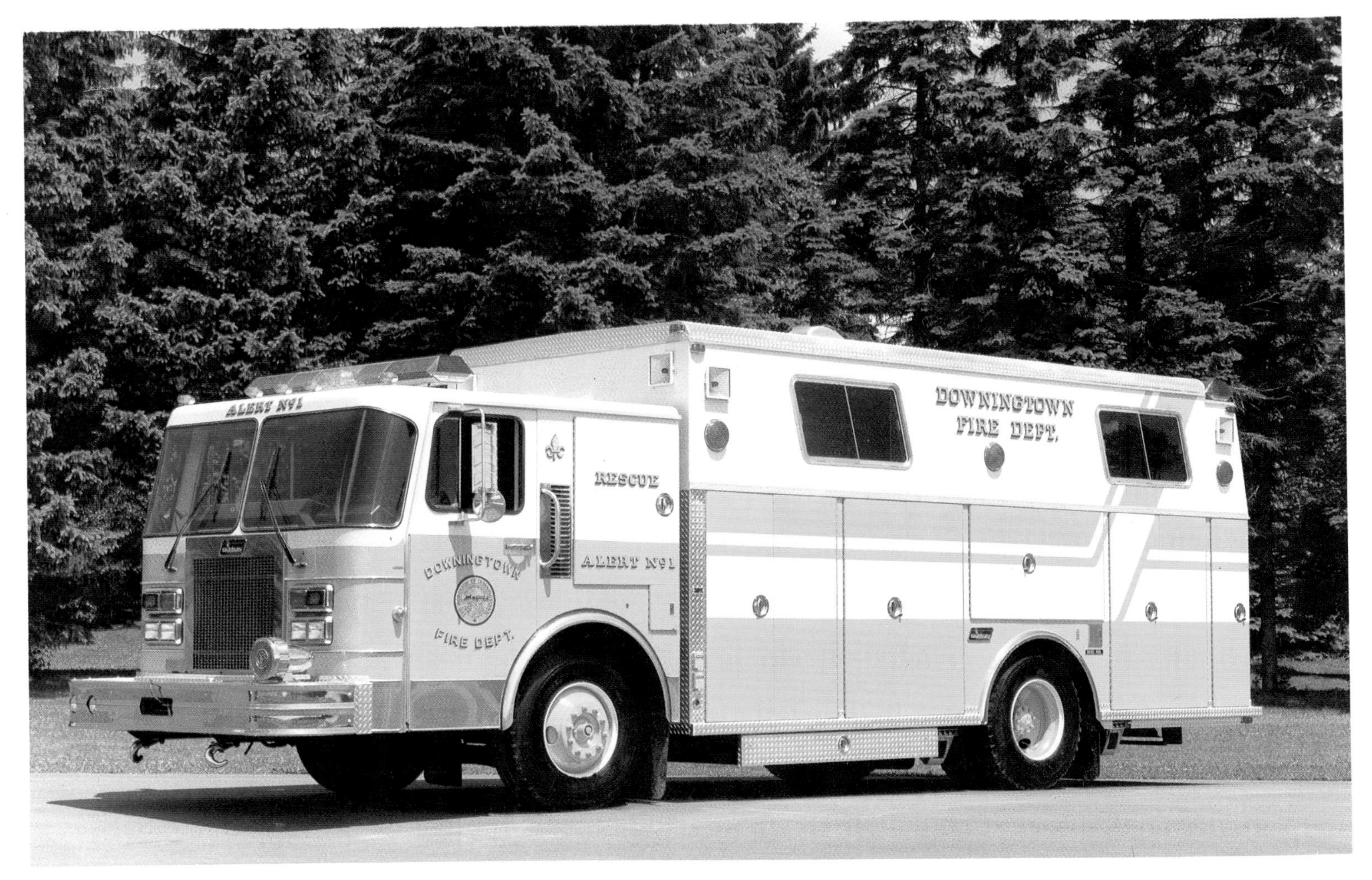
ALERT No1
DOWNINGTOWN
FIRE DEPT.
RESCUE
ALERT No1
DOWNINGTOWN
FIRE DEPT.

6-26

6-27

TOWAMENSING VOL. FIRE CO. NO. 1
TRACHSVILLE, PA.
TOWAMENSING TWP.
VOL. FIRE CO.
NO. 1
TANKER
PUMPER
932
MACK

stylish three-piece radiator grille. The firm used the same grille design for nearly the next 20 years, which makes Ward LaFrance rigs from the period very easy to spot. Another typical Ward LaFrance detail is a post-mounted siren ahead of the driver's-side door. In 1949 Ward LaFrance began offering a new production model called the Elmira Eagle. Available in 500- and 750-gpm versions, this new model had an optional three-man closed cab. It was offered throughout the 1950s, and hundreds were sold.

Ward LaFrance began offering a new, smaller production pumper called the Fireball Special in 1954. A feature of this model was the 500-gallon booster tank, which made it useful for small towns and rural areas.

The John Bean Division of the Food Machinery Corporation (better known today as FMC) began offering high-pressure fog trucks in the late 1940s. High-pressure fog was a very effective firefighting technique that had been developed during the war, primarily to fight shipboard fires. Later the method was applied to civilian firefighting and proved extremely effective. The fog could often knock down the fire by itself. In other cases, the fog provided a "curtain" that protected the firefighters as they maneuvered into position to attack the flames; moving fog curtains allowed the firefighters to advance more safely into burning buildings. A major advantage of fog is that it uses a relatively small amount of water, thus minimizing water damage and flooding and placing less strain on the supply system. A typical fog system might put out only 60 gpm, but at a pressure of 850 psi. Beginning in the late 1940s, FMC delivered its high-pressure fog apparatus on a variety of commercial chassis. The firm also produced combination pumpers featuring high-pressure fog guns, usually on GMC commercial chassis.

THE SNORKEL

One day in 1958, Chicago Fire Commissioner Robert Quinn happened to notice some city workers cleaning an overhead sign. The workers were using a cherrypicker—an articulated boom with a bucket large enough to hold two men, mounted on the back of a small truck. Quinn was struck by the ease with which the workers could maneuver into position, and had the original and brilliant idea of using the same concept for firefighting. Quinn's insight was sharpened by the knowledge that the department's three antiquated water towers needed to be replaced, but no manufacturer was making them any more.

Because the cherrypicker concept itself had been around since 1951, the Chicago department's shop was able to purchase a 50-foot articulated boom with basket from the Pitman Manufacturing Company in Missouri and mount it on an ordinary truck chassis. A length of hose was placed alongside the boom and attached to a 2-inch nozzle in the basket; the final result produced 1,200 gpm at up to 100 psi. This was the first elevating platform in the fire service, but it has gone down in history by the name of Quinn's Snorkel. The legendary source of the name comes from the device's first use. In October 1958, a month after it was placed in service, the rig was called out for a four-

6-26
A Cardinal series 18-foot rescue vehicle custom-built by Saulsbury. The chassis is a Ford L-8000; the engine is a 250-hp Caterpillar diesel.

6-27
A Mack MC tilt-cab chassis was used on this custom-built 750-gpm pumper/tanker from Saulsbury. The fiberglass booster tank holds 2,500 gallons.

6-6

6-7

alarm lumberyard fire. It was spectacularly effective, dousing the blaze in short order. Speaking with reporters afterward, the firefighter operating the basket hose dubbed the rig "Commissioner Quinn's snorkel," because he had been under the water from the streams of other rigs the entire time. The name stuck, and Commissioner Quinn was quickly dubbed Snorkel Bob.

The unusual new rig received wide publicity and praise after it was used to fight the disastrous Our Lady of Angels school fire on 1 December, 1958. This fire, which happened in an unsprinklered building with open stairwells, took the lives of 92 children and three teachers. The snorkel featured in the photographs of the disaster that appeared on the front pages of newspapers across the nation.

The Pitman Company quickly capitalized on its head start in the elevating platform business, but other firms soon joined the competition. The Snorkel Fire Equipment Company was formed in St. Joseph, Missouri. Existing firms such as Mack, American LaFrance, Ward LaFrance, and Sutphen all began offering elevating platforms in various configurations. In addition to standard elevating platforms on articulated booms, telescoping booms for constricted spaces were designed. The Snorkel Company came up with a combination water tower and aerial ladder on a telescoping boom. These models, called the Squrt and the Telesqurt, could be mounted on pumpers, giving a new twist to the combination concept.

AERIAL LADDERS

Wooden aerial ladders and traditional city service ladder trucks were the norm in many departments before the war, even though Seagrave had introduced all-steel aerials in 1935 and Pirsch had introduced all-aluminum aerials in 1936. In the late 1940s and early 1950s, however, new designs and techniques were making metal aerials increasingly popular, and city service ladder trucks were on their way toward obsolescence. The basic concepts of the four-wheel chassis carrying a multiple-section, hydraulically operated aerial ladder driven by a power take-off from the transmission, and stabilized by extendable jacks, were well in place by the late 1940s. In 1948 the New York City fire department purchased its first hydraulically operated, all-metal aerials. These were tractor-drawn rigs provided by American LaFrance, Pirsch, and Seagrave. Surprisingly, however, in 1955 NYFD took delivery of 25 tractor-drawn, spring-raised wooden aerials, provided by FWD.

Mack began offering new aerial ladder trucks in 1948, using all-metal ladders provided by Maxim. In 1952 Maxim obtained North American rights to the German Magirus aerial ladder and began offering it in a rear-mounted configuration. Inexplicably slow to catch on, since these maneuverable rigs were actually not much bigger than a standard pumper and were thus convenient for use on crowded streets, rear-mounted aerials did not start becoming the norm until the 1960s. When Maxim became a part of Seagrave in 1956, the Magirus designs continued to be offered through Seagrave and also through Mack.

6-6
Fully enclosed and very streamlined for its time, this 1948 Mack rescue vehicle was used by FDNY Rescue Company No. 1.

6-7
This 1947 Mack Type 85 pumped 650 gpm for NYFD Engine Company No. 23.

FWD began offering another European aerial ladder, made by the Dutch company Geesink, in 1956. These rear-mounted ladders came in lengths as long as 105 feet, but were generally more compact than other designs.

THE SIXTIES AND DIESELIZATION

The diesel engine was patented in 1892 by a German engineer named Rudolf Diesel. While diesel engines operate on the same four-cycle principle as gasoline engines, the ignition phase is achieved by compressing the air in the cylinder (thus raising its temperature) and then injecting fuel into the cylinder; no spark plugs are needed. The fuel is ignited by the high temperature of the hot air; it burns and forces the piston downward to create the power stroke. Finally, another upward stroke of the piston forces out the exhaust and the cycle begins again. Diesel engines offer several advantages: They are simple, sturdy, and reliable; they operate on cheaper fuel; and they are more efficient for powering large, heavy apparatus such as aerial ladders and tankers. American LaFrance was the first to offer diesel engines on fire apparatus, and within a few years diesel engines were the norm for all manufacturers (although smaller rigs built on commercial chassis continue to be available with gasoline engines).

By 1961 American LaFrance was no longer offering its own V-12 engine, primarily because simpler and cheaper six-cylinder engines that delivered the same horsepower were now being made by competing companies such as Waukesha. In 1965 American LaFrance again pioneered by offering diesel engines.

In 1966 American LaFrance had become a division of A-T-O Inc., formerly known as "Automatic" Sprinkler. In 1963, the near-bankrupt "Automatic" Sprinkler Company had been purchased by a group of investors led by Harry E. Figgie, Jr. Thus American LaFrance became, and still remains, part of Figgie International.

In 1967 the Mack C Series was improved and renamed the CF Series. Also in 1967, Mack replaced the old B Model with a new conventional chassis called the R Model.

In 1965 Mack delivered to New York City one of the most unusual and famous pieces of fire apparatus ever built. Dubbed the Super Pumper, this feat of engineering was an incredibly powerful pumping engine that was designed to take the place of ten ordinary pumpers. The idea was to have one rig that could pump massive amounts of water onto the sort of major fire that might be expected in New York's skyscraper territory. Hauled by a Mack F Series diesel tractor, the Super Pumper had a six-stage centrifugal pump capable of delivering an incredible 8,800 gpm at 350 psi, drawing from as many as eight hydrants. The Super Pumper was accompanied by the Super Tender, which featured a an 8-inch, 10,000-gpm water cannon and carried 2,000 feet of 4 1/2-inch hose. In addition, there were three C Series Satellite Tenders, each with a 4,000-gpm deck turret. The amazing Super Pumper remained in service until

6-28
A Walter R 2000 rapid-intervention vehicle. This vehicle carries a 75-gallon foam tank, 500 pounds of dry chemical, and a 600-gallon water tank. The roof turret is remote controlled.

6-29
Walter Truck Company built this four-wheel-drive crash rescue vehicle for the US Coast Guard. The conventional design is unusual for Walter.

6-28

6-29

BOXING GUILD OF NEW YORK
AFFILIATED WITH
INTERNATIONAL BOXING GUILD
INTERNATIONAL BOXING GUILD
UNION-ALL
ONE WAY
NO PARKING
F.D.N.Y.
2
F.D.

6-8

6-9

OSHKOSH
2
OSHKOSH
M16-032

1982, when it was replaced by no fewer than six double units of 2,000-gpm pumpers and accompanying hose wagons.

One problem with cab-forward apparatus is access to the engine compartment for servicing. At first, removable panels were added inside the cab. In 1949, White Motors introduced the first tilt-cab chassis. This innovative concept allowed access to the engine by literally tilting the cab forward. The concept quickly caught on among the commercial chassis makers and the fire apparatus companies. In 1958, Mack introduced its first tilt-cab truck models, the N Series, using a chassis built by the Budd Company. This was followed in 1962 by the F Series, which remained in production until the early 1980s. Another popular Mack tilt-cab chassis was the MB series, introduced in the mid-1960s. This chassis was exceptionally maneuverable, which made it ideal for city service fire apparatus.

A significant conceptual breakthrough came from Howe in 1967. This was the raised pump-control panel. Instead of putting the pump operator's control panel in its traditional place on the right side of the truck, Howe placed the panel transversely above the back of the cab. This gave the pump operator a full view of the fireground. Despite the obvious advantages of an elevated pump panel, this concept was very slow to catch on, although more and more modern pumpers offer it as an option.

Peter Pirsch & Sons did not introduce a cab-forward model until 1962, and continued to offer its engine-ahead models until into the 1970s. Commercial models were also made on both conventional and cab-forward chassis, according to the customer's spec. Pirsch was and is today best known for manufacturing aerial ladders.

THE SEVENTIES

One aspect of the late 1960s and early 1970s that was difficult for the fire service was the elimination of racism. All too many departments had made it difficult or impossible for blacks to enter, and had sometimes left black neighborhoods dangerously underserved. A changing social climate and strict new laws meant new opportunities for African-Americans, and many took advantage of preferential hiring to become firefighters. While controversial at the time, and not at all helped by the attitude of many ghetto dwellers toward the fire service, preferential hiring helped establish racial equality in the fire house. Since then, although unfortunate pockets of prejudice still linger, black firefighters have become accepted for their own merits. The determined few who broke the color bar paved the way for later blacks, Hispanics, and women.

The American LaFrance Series 900 models were joined in 1970 by the Series 1000 models, featuring standard diesel power. In 1973 the Century Series of custom apparatus was introduced. Incorporating many of the standard features of the Series 1000 rigs, the Century Series offered a redesigned cab and improved visibility. In 1975 ALF began offering the Stinger line of quick-attack pumpers. These rigs were built on Chevrolet or Dodge 4 x 4 chassis; they were

6-8
In 1947 the NYFD ordered 20 of the new American LaFrance Series 700 open-cab pumpers. These rigs had 750-gpm pumps.

6-9
Oshkosh delivered this triple combination in 1957. Note the angled intake valve by the bell on the left front bumper and the use of compartments built into the rear bumpers.

6-35

equipped with 250-gpm pumps and 250-gallon stainless steel water tanks. Another commercial pumper line, the Spartan, was introduced in 1976. This budget-priced rig had a Pioneer-type cab with seating for five crew members. Options included a choice of diesel or gasoline engine, 500- or 750-gpm water tank, and pump capacity ranging from 750 to 1,250 gpm. Other ALF introductions in the 1970s included the Challenger/Conquest line in 1978, the 100-foot Water Chief aerial ladder in 1976, and the 75-foot Water Chief aerial ladder in 1979.

In addition to chassis sales for commercial rigs, Mack began offering its own fire engine built on the MB starting in 1973. It was a popular offering, and more than 300 had been sold by the time the model was phased out in 1978. Replacing the MB Series were two new chassis, the MC and MR Series. These too are highly maneuverable and available in several different configurations, making them an excellent foundation for fire apparatus assemblies. The MR Series was designed for heavy-duty use such as tankers, while the MC Series was smaller and meant for lighter work. The MC Series is perfectly capable, however, of such strenuous work as carrying aerial platforms and serving as a tanker. Today these chassis are frequently specified by departments buying custom rigs from body manufacturers.

6-36

Mack continues to be a major supplier to the New York City Fire Department. In 1974 the department took delivery of its thousandth gasoline pumper—a 1,000-gpm Mack CF Series diesel. In 1979 Mack delivered 40 1,000-gpm pumpers to the FDNY, and another 74 were delivered in 1980.

Starting in the early 1970s, Ward LaFrance offered its equipment in a lime-yellow color, claiming that this improves visibility and safety. The claim remains controversial. Here, as in many others aspects of the fire service, tradition gives way to the evidence only slowly.

The Pierce Manufacturing Company, based in Appleton, Wisconsin, had long been a builder for other apparatus manufacturers. In 1968 the firm began successfully marketing its own line of apparatus on commercial chassis, called the Suburban. The Power Chief line was introduced in 1970.

Mergers were part of the 1970s. By the middle of the decade, Howe Fire Apparatus had merged with another well-known manufacturer, Oren Roanoke Corp., and in turn was merged into Grumman Emergency Products in Roanoke, Virginia. The new company produces pumper, aerials, tankers, and minipumpers. The Firecat pumper was offered starting in 1978; the inexpensive Wildcat pumper on Ford commercial chassis followed shortly thereafter.

6-35
Pierce introduced the cab-over-engine tilt-cab Dash pumper in 1984. This rig carries a crew of four.

6-36
The Pierce Dash pumper can carry a water tank of up to 1,500 gallons on a single rear axle.

OSHKOSH
OSHKOSH
CALEDONIA
LEWISTON · PACIFIC
FT. WINNEBAGO
4 TOWNSHIP F.D.

6-10

6-11

OSHKOSH
OSHKOSH
NO. 2
OSHKOSH FIRE DEPARTMENT

CRASH VEHICLES

Commercial and civilian aviation expanded rapidly after World War II. Federal regulations and also common sense called for thorough fire protection of airports. Over the years apparatus manufacturers learned a great deal about air crash rescue and firefighting from the military, particularly from the experience gained on aircraft carriers and in the Air Force. American LaFrance, for example, built more than a thousand Type 0-11 and 0-11A crash trucks for the Air Force from 1950 to 1953. The expertise acquired here was also applied to civilian use. Special vehicles to deliver foam, Purple K powder, and other fire-suppressing chemicals have been developed. Many of the behemoths found at major airports are designed for worst-case firefighting: the catastrophic crash and subsequent burning of a jumbo jet. To that end, the apparatus is capable of traveling at high speed over rough terrain and then delivering huge amounts of chemical from a distance and while in motion. During the initial attack, the firefighters remain protected within the cab. Fortunately, firefighters at an airport are rarely called upon to use their heaviest equipment, and spend much of their time on inspection and fire-prevention duty, as well as training and extensive preplanning for emergencies. The potential for fire at an airport is high, but most calls are routine and can be handled with conventional equipment.

THE EIGHTIES

In the 1980s the venerable American LaFrance headquarters in snowy Elmira, New York, were moved to a modern factory in more temperate Bluefield, Virginia. Despite the upheaval, ALF delivered 80 1,000-gpm pumpers to New York City in 1981, nine more in 1982, and 14 in 1984. Current offerings from ALF include the Pioneer line of budget-priced commercial pumpers, the Pacemaker commercial pumpers, and the Century 2000 custom line.

The Pioneer line from American LaFrance offers a 1,250-gpm Twinflow pump, a four-man tilt cab, and a 240-hp Cummins diesel engine. The Pacemaker line is built on a Pemfab chassis with tilt cab. A typical pumper in this line has a 1,500-gpm Hale pump, a 500-gallon fiberglass water tank, a 300- or 350-hp diesel engine, and seats six. This chassis can also be used to carry Squrt and Telesqurt articulating Snorkel units.

The Century 2000 custom line is the premier offering from American LaFrance. Featuring stainless-steel modular construction for easy maintenance, repair, and replacement, these units are powered by 350-hp Detroit Diesel V-6 Turbo diesel engines. The pump is a 1,500-gpm Twinflow; a foam pumping system is standard equipment. Cab options include four-door, seven-man or two-door, five man styles. The wheelbase on these rigs is only 166 inches, but the compartment space is a roomy 167 cubic feet.

Aerial ladders currently available from American LaFrance include the Water Chief and Ladder Chief designs, all available on the Pacemaker and Century 2000 chassis. The 100-foot rear-mount Ladder Chief and Water Chief re-

6-10
A sturdy-looking 1950 Oshkosh triple combination delivered in Wisconsin in 1950.

6-11
This streamlined but rugged-looking fire engine was used by the Oshkosh, Wisconsin, Fire Department. Naturally, it was built by Oshkosh in 1945.

6-37

6-38

6-37
This Pierce Arrow combination has black vinyl pump panels, a Pierce trademark.

6-38
Aerial ladders from Pierce are available in 55-, 75-, and 105-foot lengths. The 75-foot ladder shown here is mounted on a Dash pumper.

6-39
A raised crew cab roof that can serve as a mobile command station is featured in this Lance pumper from Pierce. The extended-door cab can seat ten crew members.

6-40
A Lance pumper from Pierce, introduced in 1985. The canopy cab in this model seats six.

6-41
All-aluminum cab construction and a Waterous pump are features in this Pierce Arrow model. The Arrow was introduced in 1984.

6-42
The Oshkosh T-3000 crash/fire/rescue vehicle weighs 69,000 pounds. It can accelerate from 0 to 50 mph in 40 seconds, and attains a top speed of 65 mph.

6-43
This monstrous rig from Oshkosh is a model P-15, designed for the United States Air Force. It holds 6,000 gallons of water.

6-44
Oshkosh built this P-19 rapid intervention vehicle for the United States Marine Corps. The water tank holds 1,000 gallons, and the foam tank holds 130 gallons.

6-39

6-40

6-41

6-42

6-43

6-44

BLAUVELT VOL.
2
FIRE CO.

6-12

6-13

DELMAR
ENGINE
2
DELMAR
2
FIRE DEPT.
Mack

quire single rear axles and offer low travel height. The 100-foot Quint rear-mount Ladder Chief and Water Chief need double rear axles on the chassis. The 75-foot Water Chief aerial ladder is made entirely of steel. It offers a 500-pound tip load and 180 degree nozzle sidesweep.

Because Mack Trucks had come to be owned in part by the Renault company of France, in 1980 the firm began offering a new chassis with a Renault engine. Called the MS Series, this model has had a checkered career, but it is still available and has seen some success as chassis for fire rigs. By the mid-1980s Mack was producing fire apparatus in conjunction with Ward LaFrance. The Mack/Ward combination seems to be a good one and New York City continues to purchase them, starting with 25 1,000-gpm pumpers delivered in 1985. As of 1989, well over 100 Mack/Ward pumpers are in active service with FDNY.

In the late 1980s Pierce has introduced the Javelin line of rear-engined, front-wheel-drive fire engines. The engine is located just forward of the rear axle. This makes the cab not only more spacious (it can accommodate up to ten people), but also considerably quieter. The more traditional Dash line from Pierce uses a hydraulic tilt cab.

FMC in the 1980s began offering a much wider range of highly customized fire vehicles, some built on custom chassis. The firm made a deliberate decision to focus on the rural market and offer a wider variety of options and new products. In 1985 the firm moved from its antiquated, 70-year-old plant in Tipton, Indiana, to a new facility in Orlando, Florida. Current offerings include the Starfire and Sentinel lines.

In the mid-1970s a new apparatus firm called Emergency One entered the market. A division of the well-known Federal Signal Company, E-One has grown rapidly and now holds a prominent place in the industry. E-One operates from a state-of-the-art facility in Ocala, Florida. The firm offers, custom, commercial, and quick-attack pumpers, tankers, aerials, elevating platforms, and a variety of rescue and hazmat vehicles. E-One pioneered in the use of heavy-duty extruded aluminum and aluminum plate for body construction. Strong, rigid, and light (about one-third lighter than steel), aluminum is also far less subject to corrosion—all qualities that make it particularly well suited for fire apparatus.

Aerial ladders and elevating platforms remain an important part of the firefighter's arsenal. Strict safety and training requirements combined with technological improvements in the 1970s and 1980s to make these rigs more effective. Problems of stability have caused some rigs to collapse or overturn. This has been significantly reduced by better outriggers and leveling systems, lower centers of gravity, and improved chassis design to increase rigidity and reduce torque. A good example is the 102-foot elevating Aerialcat platform from Grumman Emergency Products in Roanoke, Virginia. This rig extends to a full 102 feet and through a full 360 degree rotation. The platform nozzles have a flow capability of 2,000 gpm. The Grumman Aerialcat 121-foot aerial ladder

6-12
Built by Saulsbury, this Sierra series custom mini-pumper has a 450-gpm pump, a 250-gallon booster tank, and an all-aluminum body. It is built on a Ford F-350 four-wheel-drive chassis and has a 170-hp diesel engine.

6-13
Saulsbury custom-built this 1,500-gpm apparatus on a Mack CF chassis. This rig carries a foam system, a 750-gallon booster tank, and 1,000-watt quartz floodlights. The propulsion is provided by a Mack 350-hp diesel engine.

produces 1,500 gpm, and can carry its full load safely at any extension and elevation angle. Torque problems on aerials from Emergency One have been reduced by the use of an integrated torque box chassis that provides an extremely rigid frame. This has enabled E-One to build a 135-foot aerial ladder—the tallest in North America.

6-45

6-46

6-47

Not every department needs a 135-foot aerial, however, and smaller sizes are more usual. Good examples among elevating platforms are the 75-foot and 95-foot Aerialscopes from Baker Equipment Engineering Company. Sutphen Corporation continues its long tradition of fine aerial ladders, offering ladders as long as 104 feet. Other well-known manufacturers include Snorkel (now a Figgie International Company), Seagrave Fire Apparatus, and Simon Ladder Towers Inc.

In the good old days, firemen were called out when there was a fire—or maybe a cat stuck in a tree. The fires themselves usually involved wooden structures, and the hazards involved were fairly predictable. But as daily life becomes more complex and technological, firemen are increasingly called out to handle problems such as leaking gasoline trucks, chemical spills, and fires in electrical transformers. Even routine structure fires are more hazardous, since the contents of the building and even the structure itself often contain plastics and other products that give off dangerous fumes when burned. Advances in self-contained breathing apparatus (SCBA) have helped make fighting these fires much safer, but the problem of dealing with hazardous materials is permanent.

Ongoing hazardous materials (hazmat) problems have led to ongoing research and innovation in apparatus design since the mid-1970s. Today's hazmat vehicle is often mounted on a commercial truck or van chassis. The essential requirement is lots of easily accessible storage for the wide range of equipment carried. On the larger trucks, this can include air bottles, compressors, winches, generators, lights, and decontamination gear. The most up-to-date (and expensive) hazmat trucks are equipped with advanced communications gear and a computer system with a complete hazardous materials database.

An interesting innovation that came into its own in the 1980s is the attack or mini pumper. These smaller units are in some ways a throwback to the heyday of the chemical engine, since their purpose is similar: a quick response to a small fire. Versatile, maneuverable, and requiring a smaller crew complement, these rigs today are often the first responders to a call. As a rule, attack pumpers are mounted on short wheelbase commercial truck chassis, often with four-wheel drive. They usually carry a smaller pump designed to put out 250 to 500 gpm, although some attack pumpers can produce as much as 1,000 gpm. Most units also carry tanks holding anywhere from 100 to 500 gallons.

6-45
The DA-1500 aircraft vehicle and firefighting vehicle from Oshkosh is articulated for faster speeds and greater mobility. The eight driving wheels give this vehicle the ability to move rapidly over soft ground where more conventional vehicles would bog down.

6-46
The T-1500 crash/fire/rescue vehicle from Oshkosh can discharge foam at 125 gpm at 250 psi. Like all modern crash vehicles, it offers side discharge outlets, a roof turret with remote control, and undertruck nozzles.

6-47
This Starfire pumper from FMC is built on a four-door Ford Chassis with on-board lighting, a 1,000-gallon water tank, and a 1,250-gpm pump.

6-17

THE FUTURE

In many ways, the fundamental development of the fire engine has gone as far as it can. Although there have been many significant improvements and refinements in the basic fire engine design over the decades, conceptually not that much has changed since the introduction of cab-forward models in the early 1950s. Modern apparatus builders are far from complacent, however, and efforts to improve performance and firefighter safety and comfort are constant, for their own sake and also in compliance with strict federal regulations. One area where attention is currently focused is noise reduction. Too many firefighters are developing hearing-loss problems as a result of riding in noisy cabs while listening to screaming sirens and braying horns. Another significant area of design innovation is in hazmat vehicles. As the fire service is increasingly called upon to handle complex and dangerous problems with hazardous materials, the manufacturers of these vehicles are responding with new design concepts and specialized equipment. Constant improvements in communications equipment, SCBA apparatus, and personal safety gear mean that firefighters today are less likely to be injured on the job and are less likely to develop job-related illnesses and disabilities—although the dangers remain high. Despite every improvement, all firefighters are potentially at risk every time they answer an alarm. In the end, courage and dedication remain the firefighter's most important equipment.

6-14
An International-Navistar chassis is the basis for this custom-built Saulsbury pumper/tanker. The front-mount pump moves 1,000 gpm; the fiberglass booster tank carries 2,000 gallons. A 300-hp Cummins diesel with 10-speed manual transmission powers this rig.

6-17
Pump panels today feature solid-state electronics. This raised panel is on a Saulsbury Five Star pumper/rescue rig. Although it can't be distinguished here, the pump panel is color-coded for easier operation.

6-14

6-48
Stainless-steel construction is featured in this 3,000-gallon Supertanker from FMC. Its fully enclosed cab, intersection lights, and high-visibility paint increase firefighter safety.

6-49
This custom FMC Sentinel pumper offers top-mount pump controls. It is painted an unusual but very attractive navy blue.

6-50
A fully enclosed FMC Sentinel with side-mount pump panel, reflective stripe, and advanced lighting.

6-51
Built on a fully enclosed Mack chassis, this FMC Sentinel has a hydraulic ladder rack.

6-52
This FMC Starfire tanker is constructed on a Ford chassis. It has a 1,250-gallon water tank and a 250-gpm pump.

6-53
An FMC Sentinel pumper/tanker. This rig carries 3,000 gallons of water.

6-54
This custom FMC Sentinel has an ergonomically designed pump panel and pump outlets on the front bumper.

6-48

6-49

6-50

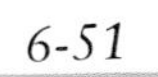

6-51

6-52

WESTERN STAR
TANKER 35
SOLOMON RUN
VOL.
FIRE DEPT.
FMC

6-53

JACKSONVILLE
471
VOL. FIRE CO.
BALTIMORE COUNTY
FMC

6-54

LAKE SAINT LOUIS
FIRE PROT. DIST.
FIRE RESCUE

6-15

6-16

FIRE RESCUE
LAKE SAINT LOUIS
FIRE PROT. DIST.

6-18

6-19

6-15
A contemporary design is shown in this Five Star pumper/rescue apparatus from Saulsbury. The four-door cab holds ten fire-fighters. Note the raised pump panel at the rear by the deluge gun. This gives the pump operator an unobstructed view of the fire scene.

6-16
A rear view of a Five Star pumper/rescue apparatus from Saulsbury. This rig offers a very generous 110 cubic feet of hose bed capacity.

6-18
An aircraft crash truck built by Walter Truck Company in 1967. Walter is known for four-wheel-drive vehicles featuring worm gears in the differential.

6-19
Tires this size take work to clean. This Walter Truck Company crash rescue rig was built for the Federal Aviation Agency for use at Dulles International Airport in Washington, DC in the 1970s.

6-55

6-56

6-55
An FMC Telesqurt is mounted on a Ford chassis. The Telesqurt has a boom, ladder, and water pipe.

6-56
This custom FMC Sentinel top-mount pumper has vertical crosslays for the hose.

6-57
Does lime green really enhance visibility? Today more and more departments think it does, as shown by this handsome FMC Starfire, built on a GMC Topkick chassis.

6-58
A hydraulic ladder rack on this custom FMC Sentinel with top-mount pump controls allows more compartment space.

6-59
On this custom FMC pumper/tanker, the engine is mounted behind the driver for more cab space.

6-60
FMC's Starfire line is designed for rural fire departments. This rig is on a Ford chassis. The air horns and speaker are mounted in the front bumper for lower noise levels in the cab.

6-61
On this fully enclosed FMC pumper/tanker the hose trays are in front of the top-mount pump controls for easier access.

6-62
This FMC Starfire on an International Harvester 4 x 4 chassis is equipped for rapid response.

6-57

6-58

6-59

6-60

6-61

6-62

6-30

6-20

6-31

6-21

6-20
City, county, and federal apparatus from all over California await assignment to duty at the massive Sierra Fire on the Sierra and Stanislaus National Forests in 1987.

6-21
Pierce recently delivered this pumper to Bayside, New York. Note the black pump panel.

6-30
This fierce-looking rig was developed by Lockheed for the Bureau of Land Management in 1974. Called the Dragon Wagon, it is designed to allow a three-man crew to fight forest and brush fires from the safety of the cab.

6-31
A Walter Truck Company foam tender, built for the City of Richmond Municipal Airport. This rig has two under-truck nozzles and a remote-controlled roof turret.

6-63
This fully enclosed custom FMC Starfire pumper offers advanced lighting, a front water intake, and air conditioning.

6-64
The Bureau of Land Management owns this 1943 International 2 1/2-ton, 6 x 6, 800-gallon rig.

6-65
A 1986 International "heavy foam" fire engine in action at the Yellowstone National Park fire in 1988. This rig, owned by the Bureau of Land Management, has a capacity of 1,500 gallons.

6-66
The Bureau of Land Management, which is responsible for fighting fires in national parks and forests, owns this 1986 "heavy Chevy." It has a capacity of 700 gallons; gross vehicle weight is 26,000 pounds.

6-67
The Bureau of Land Management assembles vehicles from commercial components and provides them to its "customers" around the nation.

6-68
Hose is heaped high in the back of this International Harvester 4 x 4 run by the Bureau of Land Management. This photo was taken at the devastating Yellowstone fire of 1988.

6-69
This 1957 Mack combination pumper is a Model B. This model would soon be replaced by the cab-forward Model C.

6-70
Riverdale, Maryland, runs this 1972 Mack 1,250-gpm pumper.

6-63

6-64

6-65

6-66

6-67

6-68

6-69

6-70

CHICAGO FIRE DEPARTMENT
CHICAGO O'HARE
INTERNATIONAL AIRPORT

6-32

6-33

FIRE DEPARTMENT OF
MEMPHIS
METROPOLITAN AIRPORT
A2

6-32
The Chicago Fire Department uses this Walter Truck Company foam tender at O'Hare International Airport, the busiest airport in the world. This vehicle carries a 220-gallon foam tank.

6-33
A twin-turreted crash vehicle built by Walter Truck Company in the 1960s.

6-34
This foam tender from Walter Truck Company has twin turrets. The reel for one of the two hand lines can be seen in front of the rear wheel.

6-115
Post-war demand made the 500-gpm Mack Type 45 pumper very popular. This photo was taken around 1946.

6-34

6-115

6-71
A lime-green paint job adorns this 1973 Mack pumper.

6-72
This 1971 Ward LaFrance Ambassador pumps 1,000 gpm. It belongs to the Norfolk, Virginia, Fire Department.

6-73
Arcadia, Maryland runs this pumper built on a 1972 Dodge commercial chassis.

6-74
This small but tough-looking combination pumper was built by FWD in 1962.

6-75
The John Bean Division of FMC built this 750-gpm high-pressure unit in 1969.

6-76
A Seagrave open-cab, 750-gpm pumper from the 1960s, photographed in 1971.

6-77
Baltimore, Maryland, uses a two-tone paint pattern on its rigs, as shown by this 1979 Seagrave 1,000-gpm pumper.

6-78
This big 1971 Mack fire engine has a pumping capacity of 1,250 gpm.

6-71

6-72

6-81

6-82

6-83

6-84

6-85

6-86

6-120

6-121

6-122

6-120
The city of Boston purchased a number of Mack cab-forward C Series pumpers in 1957. Here two new rigs demonstrate their pumping ability. Note the way the ladders are carried in a canted position on the right side.

6-121
The city of Milwaukee bought four aerial ladder trucks from Mack in 1957. All were built on the B Model chassis. Truck Co. No. 18, shown here, received a 100-foot Magirus aerial.

6-122
The enclosed cab on this Mack Model B pumper from 1958 seats four crew members, two in front and two behind the 1,250-gpm pump.

6-87
Although Mack had already introduced its cab-forward Model C line, many municipalities preferred to order the tried-and-true Model B. This aerial ladder dates from 1958.

6-89
This massive 1979 M-1500 Oshkosh crash/fire/rescue vehicle was built for the United States Air Force.

6-90
This 1947 Ahrens-Fox pumper has 4-inch suction hoses pre-attached to both intake valves.

6-91
A 1953 Seagrave aerial ladder rig. The bell mounted on the right-hand side of the cab is a Seagrave trademark.

6-92
This small 1946 Ford/Howe sports an unusual red-and-white paint scheme.

6-93
A closed-cab Series 700 pumper, built in 1949 by American LaFrance.

6-94
Baltimore runs this 1983 American LaFrance aerial ladder. Note the high-visibility paint scheme and the enclosed compartment for the tiller operator.

6-95
This 1947 Mack, loaded with equipment, looks ready for some hard action.

6-96
One of the last Ahrens-Fox front-mount piston pumpers was this one, built in 1949.

6-98
One of the last of a doomed breed is this 1950 Ahrens-Fox piston pumper, still owned by the Baldwin, New York, Fire Department. Note the racy-looking removable cab canopy.

6-87

6-89

6-90

6-91

6-92

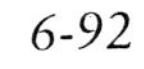

6-93

6-95

6-94

LITHOGRA H RS
PHILIP HOLZER AND ASSOCIA
YACK,N.Y.
HIGHLAND
5
HOSE CO.
GOODYEAR
GOODYEAR

6-96

BALDWIN
BALDWIN
1
FIRE DEPT
1950
AHRENS FOX
BALDWIN FIRE DEPARTMENT
"Still in Service"

6-98

Mack
GLEN ROCK, N.J.
GLEN ROCK, N.J.
FIRE DEPT.
ORGANIZED 1910
GLEN ROCK, N.J.
RADIO EQUIPPED

6-123

6-125

F.D.N.Y.
F.D. N.Y.

6-126

6-123
The first deliveries of the C Series cab-forward line from Mack began in 1957. This 1,000-gpm C-95 version, belonging to the Glen Rock, New Jersey, Fire Department, dates from 1960. NYFD fans should note the compartmented rear fenders; these are not seen on NYFD rigs from this period.

6-125
The NYFD received this Mack C-85F aerial ladder truck in 1961. It carries a six-section, 146-foot Magirus ladder.

6-126
Aerial platforms from Mack were marketed under the name Aerialscope and mounted on the C Series chassis. First offered in 1964, the Aerialscope used a 75-foot telescoping boom.

6-88

6-88
The engine is in the rear in the innovative Hush XL series of pumpers from Emergency One. The four-door cab can hold up to 12 firefighters. The rig shown here features E-One's 50-foot all-aluminum Teleboom mounted midships.

6-99
One reason Ahrens-Fox ceased production was the complexity of its piston pumper. Centrifugal pumps were easier to maintain.

6-100
A 1948 Type 85 Mack pumper in a head-on view. The famous Mack bulldog mascot is perched on the radiator.

6-101
By 1952 the Mack Type 85 had changed somewhat, particularly in the design of the radiator grille.

6-102
By 1962, the head-on view of a Mack Model B pumper looked like this.

6-103
Removable canvas cab canopies were often added to open-cab rigs. Although closed cabs were becoming more common by the time this 1952 Mack Type 85 was built, open-cab rigs were still specified by fire chiefs who didn't believe in babying fire-fighters.

6-104
A striking open-cab American LaFrance Series 700 built in 1950.

6-99

6-100

6-101

6-102

6-103

6-104

6-130

6-130
In 1967 Mack redesigned the C Series, making the cab wider and flatter. The new model was designated CF. This shot shows the CF assembly line in Allentown, Pennsylvania in August 1958.

6-97

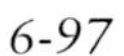

6-97
A head-on look at the 1,000-gpm piston pump of a 1949 Ahrens-Fox. By 1951, the famed Ahrens-Fox front-mounted pump, with its distinctive spherical air chamber, would no longer be in production.

6-105
This 1954 Ward LaFrance 750-gpm pumper has three-quarter doors and a removable canvas cab canopy.

6-106
An open-cab Series 700 American LaFrance aerial ladder dating from 1955.

6-107
A rear view of an American LaFrance Series 700 aerial ladder from 1955. The angle of the tractor unit gives an idea of this rig's maneuverability.

6-108
This triple combination fire engine from Ward LaFrance went into service in 1949.

6-109
This Mack Type 85 pumper was purchased by the Demarest, New Jersey Fire Department in 1954.

6-111
This sleek 1955 Ward LaFrance city service hook and ladder truck served the town of Leonia, New Jersey.

6-105

6-106

6-107

6-108

6-109

6-111

6-127

6-128

6-129

6-127
The first tilt-cab unit to enter service in the NYFD was the incredible Mack Super Pumper, delivered in 1965. The DeLaval six-stage centrifugal pump could put out an astonishing 8,800 gpm at 350 psi.

6-128
The other half of the Super Pumper System was the Super Tender, which carried an 8-inch Stang deluge gun and 2,000 feet of 4 1/2-inch hose.

6-129
The Super Pumper in action at a demonstration along the riverside shortly after delivery.

6-131
A 1,000-gpm Mack Model CF from 1971. This version has a 750-gallon booster tank and a front-mounted 5-inch suction inlet.

6-131

6-112
Pierre Thibault Trucks, Inc., of Canada built this sturdy-looking pumper/tanker on a Ford 900 chassis in 1989.

6-113
An International chassis is the basis for this modern combination rig from Pierre Thibault of Canada.

6-114
Pierre Thibault of Canada is well-known for aerial ladder trucks. This 1990 100-foot rear-mounted example offers a double-action hydraulic cylinder for extra stability.

6-136
The Century 2000 Series from American LaFrance comes with a 350-hp Detroit Diesel engine and a 1,500-gpm Twinflow two-stage pump. This pumper has a 75-foot Water Chief aerial.

6-137
American LaFrance offers the economical Pacemaker series for smaller departments. Note the location of the pump panel on a catwalk behind the cab.

6-138
The Pioneer 90 tilt-cab series from American LaFrance offers a 240-hp Detroit Diesel engine and a 1,500-gpm Twinflow pump.

6-142
A Foam Boss all-terrain crash vehicle made by CDN Research and Development Ltd. Canada and in service at Montreal International Airport. This massive rig weighs 63,000 pounds; the turret stream can reach 230 feet.

6-143
As far north as you can go and still be a volunteer fireman: Igloolik, Northwest Territories, Canada.

6-143

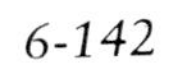

6-142

6-112

6-113

6-114

6-136

6-137

6-138

6-132

6-133

6-134

6-135

6-132
The Mack Model R replaced the old Model B in 1967. This conventional design carried a three-man crew.

6-133
This interesting Mack N Series tilt-cab unit, dating from 1962, is a quintuple rig with a 75-foot Maxim aerial ladder—probably the only such rig ever built. Note the steps built into the compartment behind the front left wheel.

6-134
Production of the Mack MC tilt-cab fire chassis began in 1979. The MC was designed for city use, and was much favored by Canadian departments.

6-135
The MS Series tilt-cab chassis from Mack was designed in Europe by Renault, which came to own a large chunk of Mack in the 1970s. The MS was introduced in 1980.

6-110
A good look at the pump panel on a Mack Type 85 from 1948. The pump produced 750 gpm.

6-139
East Pennsauken, New Jersey, is the proud owner of this Century 2000 1,500-gpm pumper from American LaFrance. The four-door cab holds seven crew members.

6-140
This beautiful blue rig was built by Sutphen Corporation. It features a Hale two-stage pump and a 75-foot telescoping water tower.

6-141
Variations on improved visibility for the commander at the fire ground is a recent trend among fire-apparatus manufacturers. This tilt-cab rig from Sutphen Corporation features maxi-vision design.

6-139

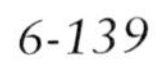

6-140

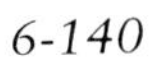

6-141

TEST PLUG
TEST PLUG
LIGHT
CLOSED
OPEN
OPEN
CLOSED
DRAIN
DRAIN
PRIMER
HOSE REEL
OTHER SIDE
HOSE REEL
THIS SIDE
AUX COOLER
CLOSED
OPEN
DRAIN
DRAIN

6-110

PHOTO CREDITS

American LaFrance: 6-136, 6-137, 6-138, 6-159; Bass Photo Collection, Indiana Historical Society Library: 5-33, 5-34; CIGNA Museum and Art Collection: 1-13, 1-14, 1-15, 2-29, 2-30, 2-31, 2-32, 2-27, 2-28, 3-63, 3-64, 3-67,3-68, 3-69, 5-137; Cincinnati Fire Museum: 3-48, 3-49, 3-50, 3-51, 3-52, 3-53, 3-54, 4-14; Colonial Williamsburg Foundation: 1-7, 1-16, 1-17, 1-18, 1-19, 1-20, 1-21; Connecticut Historical Society, Hartford: 3-19; Emergency One:, Inc. 6-88; FMC Corporation: 6-47, 6-48, 6-49, 6-50, 6-51, 6-52, 6-53, 6-54, 6-55, 6-56, 6-57, 6-58, 6-59, 6-60, 6-61, 6-62, 6-63; Forest Service Collection, National Agricultural Library: 5-4, 5-42, 5-43, 5-46, 5-47, 5-48, 6-20, 6-30, 6-64, 6-65, 6-66, 6-67, 6-68; Henry Ford Museum, Dearborn, Michigan: 3-7; Library of Congress: 1-1, 1-2, 1-3, 1-4, 1-5, 1-10, 1-11, 1-12, 2-4, 2-7, 2-8, 2-9, 2-10, 2-11, 2-12, 2-17, 2-19, 2-20, 3-5, 3-6, 3-8, 3-9, 3-10, 3-11, 3-12, 3-13, 3-14, 3-15, 3-16, 3-17, 3-22, 3-23, 3-24, 3-25, 3-26, 3-27, 3-28, 3-35, 4-2, 4-3, 4-4, 4-5, 4-6, 4-7, 4-8, 4-9, 4-10, 4-11, 4-12, 4-34, 5-2, 5-10, 5-11, 5-12, 5-13, 5-14, 5-15, 5-31, 5-32, 5-35; Lionel Martinez: 2-23, 2-24, 2-25, 2-26, 3-42, 3-46, 3-47, 3-55, 3-56, 3-57, 3-58, 3-59, 3-60, 4-25, 4-26, 4-27, 4-28, 4-29, 4-30, 4-31, 4-32, 4-33, 5-49, 5-50, 5-66, 5-67, 5-71, 5-72, 5-73, 5-74, 5-75, 5-76, 5-77, 5-78, 5-79, 5-80, 5-81, 5-82, 5-83, 5-84, 5-85, 5-86, 5-87, 5-88, 5-89, 5-90, 5-91, 5-92, 5-93, 5-94, 5-95, 5-96, 5-97, 5-98, 5-99, 5-100, 5-101, 5-102, 5-103, 5-104, 5-105, 5-106, 5-107, 5-108, 5-109, 5-110, 5-111, 5-112, 5-113, 5-114, 5-115, 5-116, 5-136, 6-96, 6-97, 6-98, 6-99, 6-100, 6-101, 6-102, 6-103, 6-104, 6-105, 6-106, 6-107, 6-108, 6-109, 6-110, 6-111; Mack Trucks Historical Museum, Inc. : 4-38, 4-39, 5-138, 5-139, 5-140, 5-141, 5-142, 5-143, 5-144, 5-145, 5-146, 5-147, 5-148, 5-149, 5-150, 5-151, 5-152, 5-153, 5-154, 6-115, 6-116, 6-117, 6-118, 6-119, 6-120, 6-121, 6-122, 6-123, 6-124, 6-125, 6-126, 6-127, 6-128, 6-129, 6-130, 6-131, 6-132, 6-133, 6-134, 6-135; Municipal Archives of the City of New York: 3-31, 3-32, 3-33, 3-34, 4-1, 4-13, 4-15, 4-17, 4-18, 4-19, 4-20, 4-21, 4-22, 4-23, 5-1, 5-3, 5--16, 5-17, 5-18, 5-19, 5-20, 5-21, 5-22, 5-23, 5-24, 5-25, 5-26, 5-27, 5-28, 5-29, 5-36, 5-38, 5-39, 5-40, 5-41, 6-1, 6-2, 6-3, 6-4, 6-5, 6-6, 6-7; New York City Fire Museum: 3-29, 3-30, 5-44, 5-45, 6-8; New - York Historical Society: 1-6, 1-8, 1-9, 2-1, 2-2, 2-3, 2-5, 2-6, 2-13, 2-14, 2-15, 2-16, 2-18, 2-21, 3-18, 3-20, 3-21, 4-16, 5-5, 5-6, 5-7, 5-8; Northwest Territories Government: 6-143; Oshkosh Company: 5-9, 5-30, 5-37, 6-9, 6-10, 6-11, 6-42, 6-43, 6-44, 6-45, 6-46; Pierce Manufacturing Company: 6-21, 6-35, 6-36, 6-37, 6-38, 6-39, 6-40, 6-41; Prometheus Colour Collection, London: Introduction photos, 2-22, 3-61, 3-43, 3-44, 3-45, 4-24, 5-51, 5-52, 5-53, 5-54, 5-55, 5-56, 5-57, 5-58, 5-59, 5-60, 5-61, 5-62, 5-63, 5-64, 5-65, 5-68, 5-69, 5-70, 5-117, 5-119, 5-119A, 5-120, 5-121, 5-122, 5-123, 5-124, 5-125, 5-126, 5-127, 5-128, 5-129, 5-130, 5-131, 5-132, 5-133, 5-134, 5-135, 6-69, 6-70, 6-71, 6-72, 6-73, 6-74, 6-75, 6-76, 6-77, 6-78, 6-79, 6-80, 6-81, 6-82, 6-83, 6-84, 6-85, 6-86, 6-87, 6-89, 6-90, 6-91, 6-92, 6-93, 6-94, 6-95; Saulsbury Company: 6-12, 6-13, 6-14, 6-15, 6-16, 6-17, 6-22, 6-23, 6-24, 6-25, 6-26, 6-27; Seneca Falls Historical Society: 3-36, 3-37, 3-38, 3-39, 3-40, 3-41; Sutphen Corporation: 3-70, 6-140, 6-141; Pierre Thibault Trucks, Inc.: 6-112, 6-113, 6-114; Transport Canada: 6-142; Walter Truck Company: 6-18, 6-19, 6-28, 6-29, 6-31, 6-32, 6-33, 6-34; Waterous Company: 3-1, 3-2, 3-3, 3-4

INDEX

Numbers in boldface indicate captions.